Springer Tracts in Modern Physics

Volume 258

For further volumes:
http://www.springer.com/series/426

Springer Tracts in Modern Physics

Springer Tracts in Modern Physics provides comprehensive and critical reviews of topics of current interest in physics. The following fields are emphasized: Elementary Particle Physics, Condensed Matter Physics, Light Matter Interaction, Atomic and Molecular Physics, Complex Systems, Fundamental Astrophysics.

Suitable reviews of other fields can also be accepted. The Editors encourage prospective authors to correspond with them in advance of submitting a manuscript. For reviews of topics belonging to the above mentioned fields, they should address the responsible Editor as listed below.

Special offer: For all clients with a print standing order we offer free access to the electronic volumes of the Series published in the current year.

Peter Schmüser · Martin Dohlus
Jörg Rossbach · Christopher Behrens

Free-Electron Lasers in the Ultraviolet and X-Ray Regime

Physical Principles, Experimental Results, Technical Realization

Second Edition

Springer

Peter Schmüser
Jörg Rossbach
University of Hamburg and DESY
Hamburg
Germany

Martin Dohlus
Christopher Behrens
Deutsches Elektronen-Synchrotron (DESY)
Hamburg
Germany

ISSN 0081-3869 ISSN 1615-0430 (electronic)
ISBN 978-3-319-04080-6 ISBN 978-3-319-04081-3 (eBook)
DOI 10.1007/978-3-319-04081-3
Springer Cham Heidelberg New York Dordrecht London

Library of Congress Control Number: 2013956525

Printed on acid-free paper

Springer is part of Springer Science+Business Media (www.springer.com)

This book is dedicated to the memory of the late Bjørn H. Wiik who was the driving force behind the superconducting linear collider project TESLA and the ultraviolet and X-ray free-electron laser projects at DESY. His determination, skillful leadership and constant encouragement were vital for the success of the superconducting cavity development and the construction of the free-electron laser FLASH

Preface to the Second Edition

In the 5 years since the appearance of this book, an enormous progress has been made in the field of free-electron lasers. The world's first hard X-ray FEL, the Linac Coherent Light Source (LCLS) at Stanford (USA), was commissioned in 2009, and the second X-ray FEL, the "Spring-8 Angstrom Compact free-electron LAser" (SACLA) at Harima (Japan), went into operation in 2012. This has been the motivation to change the title of our book to *Free-Electron Lasers in the Ultraviolet and X-Ray Regime*. Chapter 9 has been completely rewritten to describe the existing X-ray FEL facilities and report important results obtained with these machines. A second new achievement is the successful implementation of various seeding schemes at FELs working in the ultraviolet and soft X-ray spectral regions. Chapter 7 covers now Self-Amplified Spontaneous Emission and FEL seeding. Another new feature of the book is a novel derivation of the important third-order differential equation of the high-gain FEL.

The field of free-electron lasers has undergone a rapid expansion in the past few years and is still growing fast. Within the scope of this university textbook, we can cover only a small fraction of the exciting new ideas and developments. We apologize for having to omit many important results and refer to the Free-Electron Laser conferences (FEL) and the International Particle Accelerator Conferences (IPAC) as well as to the scientific journals quoted in the book for a more complete overview.

C. B. is very grateful for the hospitality extended to him at SLAC and the opportunity to participate in important scientific investigations at LCLS. We thank Hitoshi Tanaka for valuable information and comments. Particular thanks go to Zhirong Huang for his advice and numerous stimulating discussions.

Hamburg, October 2013

Peter Schmüser[1]
Martin Dohlus
Jörg Rossbach
Christopher Behrens

[1] Corresponding author Peter.Schmueser@desy.de

Preface to the First Edition

The high scientific interest in coherent X-ray light sources has stimulated worldwide efforts in developing X-ray lasers. In this book, a particularly promising approach is described, the free-electron laser (FEL), which is pursued worldwide and holds the promise to deliver ultra-bright X-ray pulses of femtosecond duration. Other types of X-ray lasers are not discussed nor do we try a comparison of the relative virtues and drawbacks of different concepts.

The book has an introductory character and is written in the style of a university textbook for the many newcomers to the field of free-electron lasers, graduate students as well as accelerator physicists, engineers, and technicians; it is not intended to be a scientific monograph for the experts in the field. Building on lectures by one of us (J. R.) at the CERN Accelerator School, and motivated by the positive response to a series of seminars on "FEL theory for pedestrians", given by P. S. within the framework of the Academic Training Program at DESY, we have aimed at presenting the theory of the low-gain and the high-gain FEL in a clear and concise mathematical language. Particular emphasis is put on explaining and justifying the assumptions and approximations that are needed to obtain the differential equations describing the FEL dynamics. Although we have tried our best to be "simple", the mathematical derivations are certainly not always as simple as one would like them to be. However, we are not aware of any easier approach to the FEL theory. Some of the more involved calculations are put into the appendices.

The starting points are the Maxwell equations and the basic elements of special relativity. We avoid the Hamiltonian formalism in the main text because many potential readers may not be familiar with this powerful formalism. A short introduction into the Hamiltonian treatment of the electron motion in an undulator magnet and its interaction with the radiation field is given in Appendix A. The FEL equations are derived in the framework of classical electrodynamics. Quantum theory is not needed to explain the theoretical basis and the functioning of presently existing or planned FEL facilities.

The differential equations describing the time evolution of the laser light wave are derived in a one-dimensional approximation and turn out to be quite powerful. In this book, they are evaluated using rather straightforward programs for computing the FEL gain curve, laser saturation, bandwidth, and other quantities of interest.

The implications and modifications of the full three-dimensional treatment are discussed.

The available experimental data on high-gain ultraviolet and soft X-ray FELs are presented but the wide field of FELs in the visible and infrared regime is not covered. We apologize for having to omit the important results obtained in this field as well as other interesting developments and refer to the literature quoted in the book and to the Free-Electron Laser (FEL) conferences and the American and European Particle Accelerator conferences (PAC, EPAC) for a complete overview over the rapidly growing FEL activities worldwide.

The International System (SI) of units is used throughout to enable the reader to obtain practical numbers from the equations in the book. Our mathematical codes (written by M. D.) are available on request. The majority of the illustrations and graphical presentations shown in the book have been prepared by us using these codes, except when otherwise noted.

We have benefited a great deal from fruitful discussions with our colleagues at DESY and other laboratories and want to thank them for their advice, in particular Evgueni Saldin, Evgeny Schneidmiller, and Mikhail Yurkov. We are very grateful to Erich Lohrmann and Sara Casalbuoni for a thorough reading of an early version of the manuscript and many valuable suggestions, and to Sven Reiche for a critical reading of the complete manuscript and his suggestions for improvement. Bernd Steffen's help with LATEX problems and editing of figures is gratefully acknowledged as well as Roxana Tarkeshian's help in checking the references. We are particularly grateful to all members of the TESLA collaboration and of the FLASH team for their invaluable contributions to the design, construction and operation of a superconducting free-electron laser in Hamburg.

Hamburg, April 2008

Peter Schmüser[2]
Martin Dohlus
Jörg Rossbach

[2] Corresponding author Peter.Schmueser@desy.de

Contents

1 Introduction 1
1.1 Overview 1
1.2 Electron Accelerators as Short-Wavelength Light Sources 3
1.3 Free-Electron Lasers and Quantum Lasers 5
1.3.1 Stimulated and Spontaneous Emission 5
1.3.2 Is the FEL Really a Laser? 7
1.3.3 Why Does the FEL Need an Undulator? 8
References 9

2 Undulator Radiation 11
2.1 Magnetic Field of a Planar Undulator 11
2.2 Electron Motion in an Undulator 12
2.2.1 Trajectory in First Order 12
2.2.2 Motion in Second Order 13
2.3 Emission of Radiation 14
2.3.1 Radiation in a Moving Coordinate System 15
2.3.2 Transformation of Radiation into Laboratory System 16
2.4 Lineshape and Spectral Energy of Undulator Radiation 19
2.5 Higher Harmonics 20
References 23

3 Low-Gain FEL Theory 25
3.1 Energy Transfer from Electron Beam to Light Wave 25
3.1.1 Qualitative Treatment 25
3.1.2 Quantitative Treatment 27
3.2 The FEL Pendulum Equations 30
3.3 Phase Space Representation and FEL Bucket 32
3.3.1 Phase Space Trajectories 32
3.3.2 Definition of the FEL Bucket 33
3.4 FEL Gain and Madey Theorem 35
3.5 Higher Harmonics and Modified Undulator Parameter 36
References 38

4 One-Dimensional Theory of the High-Gain FEL 39
4.1 General Principles of High-Power FELs 39
4.2 Microbunching 40
4.3 Basic Elements of the One-Dimensional FEL Theory 41
4.4 Electromagnetic Fields 44
4.4.1 Radiation Field 44
4.4.2 Space Charge Field 47
4.5 Corrections Due to the Longitudinal Oscillation 48
4.6 The Coupled First-Order Equations 48
4.7 The Third-Order Equation of the High-Gain FEL 52
4.8 Analytic Solution of the Third-Order Equation 57
4.9 High-Gain FEL with Helical Undulator 60
References 61

5 Applications of the High-Gain FEL Equations 63
5.1 Gain Function of the High-Gain FEL 63
5.1.1 Third-Order Equation with Energy Detuning 63
5.1.2 Short Undulator: Low-Gain Limit 64
5.1.3 Long Undulator: High-Gain Regime 65
5.2 FEL Bandwidth 65
5.2.1 Detuning Parameter 66
5.2.2 Analytical Determination of the FEL Bandwidth 68
5.3 FEL Startup by a Periodically Modulated Electron Beam Current 69
5.4 Laser Saturation 71
5.5 Linear and Nonlinear Regime of a High-Gain FEL 73
5.6 Simulation of Microbunching 75
5.6.1 Evolution of Phase Space Structures 75
5.6.2 Evolution of Phases in the FEL Gain Process 78
5.6.3 The Onset of Microbunching 82
5.6.4 Higher Harmonics in FEL Radiation 83
References 84

6 Energy Spread, Space Charge and 3D Effects 85
6.1 Increase of Gain Length by Energy Spread and Space Charge 86
6.1.1 Energy Detuning 86
6.1.2 Energy Spread and Space Charge 87
6.2 Increase of Gain Length by 3D Effects 88
6.3 Overlap Between Electron and Photon Beam 90
6.3.1 Electron Beam Focusing and Emittance 90
6.3.2 Optical Diffraction and Gain Guiding 94

6.4 Parametrization of Gain Length in an X-Ray FEL 95
6.5 FEL Radiation from Short Bunches 97
6.5.1 Velocities 97
6.5.2 Slippage Effects in Short Bunches 101
6.6 Superradiance. 105
References 105

7 Self-Amplified Spontaneous Emission and FEL Seeding 107
7.1 Computation of the SASE Process in the 1D Theory 108
7.1.1 Solution of the Third-Order Equation 108
7.1.2 Transmission Function Approach 110
7.1.3 Comparison of FEL Start-Up by Seeding or by SASE 111
7.2 Experimental Results on SASE FEL Radiation. 113
7.2.1 Detection of SASE, Exponential Gain and Saturation 113
7.2.2 Transverse Coherence of SASE Radiation 115
7.3 Statistical Properties of SASE Radiation 117
7.3.1 Simple Model of a SASE Pulse 117
7.3.2 Coherence Time. 119
7.3.3 Pulse Energy Fluctuations 120
7.4 FEL Seeding with Coherent Radiation 123
7.4.1 Seeding by Higher Harmonics of an Infrared Laser . . . 124
7.4.2 High-Gain Harmonic Generation 125
7.4.3 Echo-Enabled Harmonic Generation 128
References 130

8 The EUV and Soft X-Ray FEL in Hamburg 133
8.1 Introductory Remarks 133
8.2 Layout of the Free-Electron Laser FLASH 134
8.3 Electron Source 135
8.4 Space Charge Effects 136
8.4.1 Electric and Magnetic Forces Inside a Relativistic Bunch 136
8.4.2 Space Charge Forces in the Electron Gun 138
8.5 Superconducting Linear Accelerator 139
8.6 Bunch Compression 142
8.6.1 Principle of Longitudinal Compression 142
8.6.2 Practical Realization of Bunch Compression 143
8.6.3 Simulation of Bunch Compression 144
8.6.4 Collective Effects in the Bunch Compressors. 145
8.7 Wake Field Effects 147

8.8 Longitudinal Electron Beam Diagnostics 149
8.8.1 Transversely Deflecting Microwave Structure 149
8.8.2 Electro-Optic Detectors 150
8.8.3 Coherent Transition Radiation Spectroscopy 152
8.9 Transverse Beam Diagnostics 158
8.10 Laser-Optical Synchronization System 160
References 162

9 X-Ray Free-Electron Lasers: Technical Realization and Experimental Results 165
9.1 Photon Beam Brightness 165
9.2 The X-Ray Free-Electron Lasers LCLS and SACLA 168
9.2.1 Layout of LCLS and SACLA 168
9.2.2 Generation of Femtosecond FEL Pulses 171
9.2.3 Monochromatic X-Rays from Self-Seeding 173
9.2.4 Electron Beam Energy Loss in the Undulator Magnets 175
9.3 Undulator Radiation Background and Quantum Effects 176
9.3.1 Background from Undulator Radiation 176
9.3.2 Quantum Effects and Beam Energy Spread 178
9.4 X-Ray Beam Lines 179
References 181

10 Appendices 183
10.1 Hamiltonian Formalism 183
10.1.1 Basic Elements of the Hamiltonian Formalism 183
10.1.2 Electron Motion in a Planar Undulator 187
10.1.3 Electron Motion in a Helical Undulator 188
10.1.4 Energy Exchange Between Electron and Light Wave 190
10.2 Supplements to Chapter 4 193
10.2.1 Derivation of the Third-Order Equation Using the Vlasov Equation 193
10.2.2 Low-Gain Limit of the High-Gain FEL Theory 197
10.2.3 Beam Energy Loss During the FEL Process 199
10.2.4 Non-periodic First-Order Equations 200
10.3 Gaussian Modes of Laser Beams 204
10.3.1 Fundamental Gaussian Mode 204
10.3.2 Higher-Order Gaussian Beam Modes 207
10.4 Eigenmode Approach of 3D FEL 209
10.4.1 General Procedure 209
10.4.2 One-Dimensional Case 212
10.4.3 Cylindrical Beam with Constant Charge Density 215
10.4.4 When is the 1D Theory Applicable? 217

10.5 Statistical Methods and Tools 218
10.5.1 Current Modulation Resulting from Shot Noise 218
10.5.2 The Gamma Distribution 221
10.6 Conventions and Frequently Used Symbols 226
References 228

Index 229

Chapter 1
Introduction

1.1 Overview

The principle of the Free-Electron Laser (FEL) was invented by John Madey in 1971 [1]. The first FEL, operating in the infrared at a wavelength of 12 μm, was built at Stanford University in the 1970s by Madey and coworkers [2, 3]. For many years, FELs have played a marginal role in comparison with conventional lasers, except at microwave and infrared wavelengths. Only in recent years it has become clear that these devices are exceedingly powerful light sources in the X-ray regime [4–6]. The X-ray pulses produced in the Linac Coherent Light Source LCLS in Stanford (USA), which is described in Chap. 9, are shorter than the pulses from most other X-ray sources, and the peak brilliance is about eight orders of magnitude higher. The high pulse energy and the femtosecond duration ($1\,\text{fs} = 10^{-15}\,\text{s}$) of the X-ray pulses as well as their coherence open entirely new fields of research, for example structural analysis of individual biomolecules, which are inaccessible at the present third-generation light sources. Alternative concepts of X-ray lasers (see e.g. [7–10]) are not discussed in this book.

In this introductory chapter a comparison is made between conventional quantum lasers and free-electron lasers. Chapter 2 deals with undulator radiation which is intimately related to FEL radiation. The theory of the low-gain FEL is derived in Chap. 3. The high-gain FEL theory is treated in Chap. 4 in the one-dimensional approximation. A set of coupled first-order equations is derived as well as a third-order differential equation which permit deep insight into the physics of the FEL. The third-order equation is valid in the so-called *linear regime* of the FEL where the output field depends linearly on the input field. We often call this the *exponential-growth regime* since here the FEL power grows exponentially with the distance traveled in the undulator. The coupled first-order equations are more general and encompass in addition the *nonlinear regime* where the FEL power goes into saturation.

Numerous applications of the high-gain FEL equations follow in Chap. 5, illustrating the power of the high-gain theory. The corrections to the one-dimensional theory are discussed in Chap. 6, these comprise electron beam energy spread, space charge

P. Schmüser et al., *Free-Electron Lasers in the Ultraviolet and X-Ray Regime*,
Springer Tracts in Modern Physics 258, DOI: 10.1007/978-3-319-04081-3_1,

forces, finite electron beam radius, betatron oscillations, optical diffraction, and slippage effects in short electron bunches. Special emphasis is put on the principle of Self-Amplified Spontaneous Emission (SASE). The SASE mechanism is illustrated in Chap. 7 with numerical simulations and experimental data. Another important topic is seeding by coherent external sources. The extreme ultraviolet (EUV) and soft X-ray free-electron laser FLASH is described in Chap. 8 in some detail to give an impression of the complexity of such an accelerator-based light source. In Chap. 9 the properties of the new FELs in the hard X-ray regime and some of the excellent experimental results are discussed.

The appendices containing supplementary material are collected in Chap. 10. A short introduction into the Hamiltonian treatment of electron motion in an undulator and the coupling between electron and light wave is given in Sect. 10.1, see also Refs. [11, 12]. In Sect. 10.2 the third-order differential equation of the high-gain is derived using the conventional Vlasov-equation approach, and it is proved that the high-gain FEL theory reduces to the simpler low-gain theory if the undulator magnet is short and the growth in light intensity per undulator passage is small. The generalization of the coupled first-order equations to non-periodic cases is discussed in Sect. 10.2.4. The concepts of Gaussian beam optics are presented in Sect. 10.3 since Gaussian modes are not only important for conventional laser beams but also for FEL beams. The eigenmode approach for solving the FEL equations is explained in Sect. 10.4, and it is applied to the one-dimensional and the three-dimensional case. Section 10.5.1 deals with an important feature of SASE FELs, namely the current modulation resulting from shot noise in the electron beam. The gamma distribution describing the statistical properties of SASE FEL radiation is derived in Sect. 10.5.2. In Sect. 10.6 we summarize our conventions and list frequently used symbols, their dimension in SI units, their physical meaning, and the defining equation or the chapter where the quantity is introduced. Important formulas are put into boxes.

High-gain FELs are often considered the fourth generation of accelerator-based light sources. In contrast to existing synchrotron radiation light sources, which are mostly storage rings equipped with undulators, the FEL requirements on the electron beam quality in terms of small beam cross section, high charge density and low energy spread are so demanding that presently only linear accelerators (linacs) can be used to provide the drive beam.

Since the invention of the free-electron laser, an enormous amount of work has been done in this field. Useful reviews of the work up to 1990 can be found in the Laser Handbook, Volume 6—*Free Electron Lasers* [13]. The articles by Murphy and Pellegrini [14] and by Colson [15] present good introductions into the physics of the low-gain and the high-gain FEL. For additional reading we refer to the other articles in the handbook and to the books by Brau [16] and by Freund and Antonsen [17], and to an article by O'Shea and Freund [18]. A useful account of the progress in FEL physics and technology up to 2004 is presented in two articles by Pellegrini and Reiche [19, 20]; see also the literature quoted therein. The FEL theory is thoroughly treated on a high mathematical level in the book *The Physics of Free Electron Lasers* by Saldin, Schneidmiller and Yurkov [21]. An excellent reviews of the current status of X-ray free-electron laser theory is found in an article by Huang and Kim [22]. Very useful are also the lecture notes by Kim, Huang and Lindberg [23].

1.2 Electron Accelerators as Short-Wavelength Light Sources

In the bending magnets of a high-energy electron synchrotron or storage ring, the relativistic electrons are accelerated toward the center of the ring and emit *synchrotron radiation* tangentially to the circular orbit [24–27]. Usually different electrons in a bunch radiate independently, so the radiation is incoherent. The frequency spectrum is continuous and extends from zero to frequencies beyond the *critical frequency* ω_c

$$\omega_c = \frac{3c\gamma^3}{2R} . \tag{1.1}$$

Here R is the radius of curvature in the bending magnet, γ is the Lorentz factor

$$\boxed{\gamma = \frac{1}{\sqrt{1-(v/c)^2}} = \frac{W}{m_e c^2} ,} \tag{1.2}$$

and $W = \sqrt{p^2c^2 + m_e^2c^4}$ is the total relativistic energy of the electron.[1] The radiated power in a bending magnet of field B is

$$\boxed{P_{\text{syn}} = \frac{e^4\gamma^2B^2}{6\pi\varepsilon_0 c\, m_e^2} .} \tag{1.3}$$

Most of the power is contained inside a narrow cone of opening angle $1/\gamma$ which is centered around the instantaneous tangent to the circular orbit.

In modern synchrotron light sources the radiation used for research is produced in wiggler or undulator magnets which are periodic arrangements of many short dipole magnets of alternating polarity. The electrons move on a sinusoidal orbit through such a magnet (Fig. 1.1), the overall deflection of the beam is zero. Undulator radiation is far more useful than bending-magnet radiation because it consists of narrow spectral lines and is concentrated in a narrow angular cone along the undulator axis. The fundamental wavelength can be roughly estimated from the following consideration. Call λ_u the period of the magnet arrangement. In a coordinate system moving with the speed of the beam, the relativistic length contraction reduces the period to $\lambda_u^* = \lambda_u/\gamma$, and the electrons oscillate at a correspondingly higher frequency $\omega^* = 2\pi c/\lambda_u^*$ and emit radiation just like an oscillating dipole. For an observer in the laboratory who is looking against the electron beam the radiation appears strongly blue-shifted by the relativistic Doppler effect. The wavelength in the laboratory system is $\lambda_\ell \approx \lambda_u^*/(2\gamma) \approx \lambda_u/(2\gamma^2)$. For example, at an electron energy of 500 MeV the radiation wavelength is more than a million times shorter than the undulator period.

---

[1] The total relativistic energy of the electron is denoted by W in this book since we reserve the letter E for electric fields.

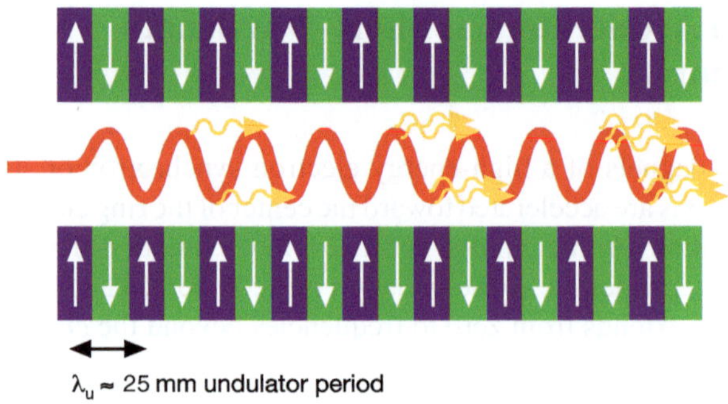

Fig. 1.1 Schematic representation of electron motion in a planar undulator and the emission of undulator radiation. In reality the alternating magnetic field and the sine-like electron orbit are orthogonal to each other but for simplicity they have been drawn in the same plane. The amplitude of the sinusoidal orbit is exaggerated, it is only a few μm.

A more accurate treatment, taking into account the sinusoidal shape of the electron trajectory and the fact that the longitudinal velocity of the electrons is lower than their total velocity, leads to the formula

$$\boxed{\lambda_\ell = \frac{\lambda_u}{2\gamma^2}\left(1 + \frac{K^2}{2}\right) \quad \text{with} \quad K = \frac{eB_0\lambda_u}{2\pi m_e c}\,.} \tag{1.4}$$

The dimensionless quantity K is called the *undulator parameter*, and B_0 is the peak magnetic field on the undulator axis. The undulator parameter is typically in the range of $1 - 3$. The proof of formula (1.4) is presented in Chap. 2. This formula, which is also valid for the FEL, refers to the fundamental wavelength $\lambda_1 \equiv \lambda_\ell$. Note that the radiation in forward direction contains odd higher harmonics with the wavelengths

$$\lambda_m = \frac{\lambda_\ell}{m}\,, \quad m = 1, 3, 5, \ldots \tag{1.5}$$

Undulator radiation has the remarkable feature that its wavelength can be varied at will, simply by changing the electron energy $W = \gamma m_e c^2$. In a limited range, the wavelength can also be varied by changing the undulator parameter.

It is interesting to note that the power radiated by a relativistic electron in an undulator is the same as that in a bending magnet with a magnetic field $B = B_0/\sqrt{2}$, however, the intensity is concentrated in a narrow spectral range. Different electrons radiate independently which means that the total energy produced by a bunch of N_e electrons is just N_e times the radiation energy of one electron. Coherent radiation with an intensity scaling quadratically with the number of electrons would be emitted if the electron bunches were shorter than the light wavelength, a condition that is practically never satisfied in the X-ray regime.

1.3 Free-Electron Lasers and Quantum Lasers

The next big improvement in the performance of accelerator-based light sources is given by the free-electron laser. The main components of an FEL are an accelerator providing a bunched relativistic electron beam and an undulator magnet. In an FEL a huge number of electrons radiate coherently because there exists a process of self-organization on the scale of the light wavelength, the so-called microbunching. The radiation power scales then quadratically with the number of these particles. For a typical number of 10^6 electrons in a microbunch the FEL will yield a million times higher light output than an undulator.

1.3.1 Stimulated and Spontaneous Emission

The word LASER is an acronym for Light Amplification by Stimulated Emission of Radiation. A conventional laser (Fig. 1.2) consists of three basic components: the laser medium with at least three energy levels, an energy pump which creates a population inversion, and an optical resonator. The axis of the optical cavity defines the direction of the photons to better than 1 mrad typically. In a mono-mode laser exactly one optical eigenmode of the cavity is excited. The photons in this mode have all the same frequency ω, the same direction, described by the wave vector $\boldsymbol{k} = (k_1, k_2, k_3)$, the same polarization and the same phase. These quantum numbers characterize a well-defined quantum state which we denote by the Dirac ket vector $|a\rangle$. Photons have spin 1 and obey the Bose-Einstein statistics; they have a strong tendency to occupy the same quantum state.

Inside the resonator there are many atoms in the excited state W_2 which can emit radiation of frequency $\omega = (W_2 - W_1)/\hbar$ by going into the ground state W_1. In the beginning of the lasing process, the number of photons in the quantum state $|a\rangle$ is zero. Call p_{spon} the probability that an atom emits its photon by spontaneous emission into this quantum state. This photon will travel back and forth between the mirrors and will remain in the cavity. However, any other photon, emitted with the same probability p_{spon} into a quantum state $|b\rangle$ with a direction different from the resonator axis, will immediately escape from the optical resonator. Therefore, the number of photons in state $|a\rangle$ increases with time. If already n photons are present in state $|a\rangle$, the probability that photon number $(n+1)$ will also go into this state is $(n+1)$ times larger than the probability p_{spon} for emission into any other state $|b\rangle$:

$$p_n = (n+1)p_{\text{spon}} = n\, p_{\text{spon}} + 1\, p_{\text{spon}} = p_{\text{stim}} + p_{\text{spon}}\,. \tag{1.6}$$

Here the term $n\, p_{\text{spon}} = p_{\text{stim}}$ stands for the stimulated emission, induced by the already existing photons in the quantum state $|a\rangle$, and the term $1\, p_{\text{spon}}$ stands for the spontaneous emission which has the same probability p_{spon} for any final state allowed by energy conservation. This equation, which can be derived in quantum field theory,

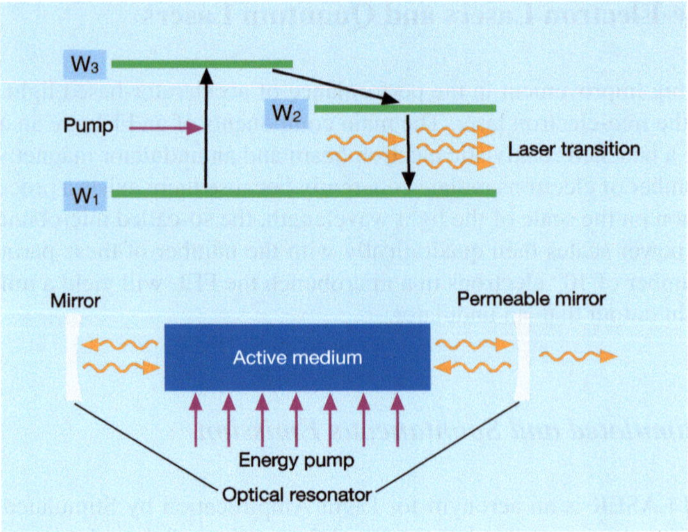

Fig. 1.2 Principle of a quantum laser where the electrons are bound to atomic, molecular or solid-state energy levels ("bound-electron laser").

is the physical basis of the laser. The lasing process starts from a stochastic process, namely spontaneous emission by the excited atoms, and the stimulated emission results in an exponential growth of the light intensity.

One can understand within the framework of quantum mechanics that the probability for stimulated emission is proportional to the number of photons already present. The usual method to compute optical transitions in quantum mechanics is by means of perturbation theory. The electron in the atom is described by a wave function obeying the Schrödinger equation. The transition from the ground state to an excited state or vice versa is caused by a perturbing Hamiltonian which is basically the potential energy of the electron in the field of an external light wave. This electromagnetic field, however, is treated as a classical quantity, using the laws of classical electrodynamics. The matrix element for the transition between two states of the atom is found to be proportional to the electric field E_0 of the light wave, and the transition probability, which can be calculated using Fermi's Golden Rule, is proportional to E_0^2. The field energy inside the volume V of the optical resonator is

$$\frac{\varepsilon_0}{2} E_0^2 \, V = n \, \hbar\omega \,, \tag{1.7}$$

where n is the number of photons in the optical cavity. Hence the probability for stimulated emission is indeed proportional to the number n of already existing photons in the quantum state $|a\rangle$.

The origin of spontaneous emission

The two radiative processes which are accessible to quantum mechanical perturbation theory are the absorption of radiation and the stimulated emission of radiation, and both have the same probability. The term p_{spon} in Eq. (1.6) corresponds to the spontaneous emission which cannot be explained in nonrelativistic quantum mechanics nor in classical electrodynamics. For a theoretical explanation of spontaneous emission not only the electron but also the radiation field must be "quantized". This leads to Quantum Electrodynamics, the quantum field theory of electromagnetic interactions. In quantum field theory, the ground state, although usually called the "vacuum", is by no means the same as the empty set in mathematics. On the contrary, the ground state is full of activity: all the time short-lived virtual photons and particle-antiparticle pairs are created and annihilated. These so-called vacuum fluctuations have a theoretically well-understood and experimentally verified influence on atomic energy levels. The spontaneous emission of radiation by an excited atom or by an electron moving through an undulator may be interpreted as emission that is stimulated by vacuum fluctuations. This interpretation has been verified in recent experiments, see the book *Exploring the Quantum* by S. Haroche and J.-M. Raimond [28].

1.3.2 Is the FEL Really a Laser?

The electrons in a conventional quantum laser are bound to atomic, molecular or solid-state energy levels, so one may call this device a *bound-electron laser*, in contrast to the free-electron laser where the electrons move in vacuum.

In a free-electron laser (Fig. 1.3) the role of the active laser medium and the energy pump are both taken over by the relativistic electron beam. An FEL operating at infrared and optical wavelengths can be equipped with an optical resonator, but this is no longer possible if the wavelength is decreased below 100 nm, because here the reflectivity of metals and other mirror coatings drops quickly to zero at normal incidence.[2] In the extreme-ultraviolet and X-ray regime a large laser gain has to be achieved in a single passage of a very long undulator magnet. The principle of Self-Amplified Spontaneous Emission (SASE) allows to realize high-gain FELs at these short wavelengths; seeding by an external coherent source is a promising alternative.

We will see in the chapter on the low-gain FEL theory that the coupling between the electrons and an already existing light wave in the undulator is proportional to the electric field E_0 of the light wave, and the laser gain is proportional to E_0^2 and thereby to the number of photons in the light wave. Hence one is well justified to speak of *light amplification by stimulated emission of radiation* when talking about a free-electron laser. Moreover, the light emerging from an FEL has the same properties as conventional laser light: it is nearly monochromatic, polarized, extremely bright, tightly collimated, and possesses a high degree of transverse coherence.

---

[2] Recently considerable progress has been achieved with crystal optics utilizing Bragg reflection in diamond [29].

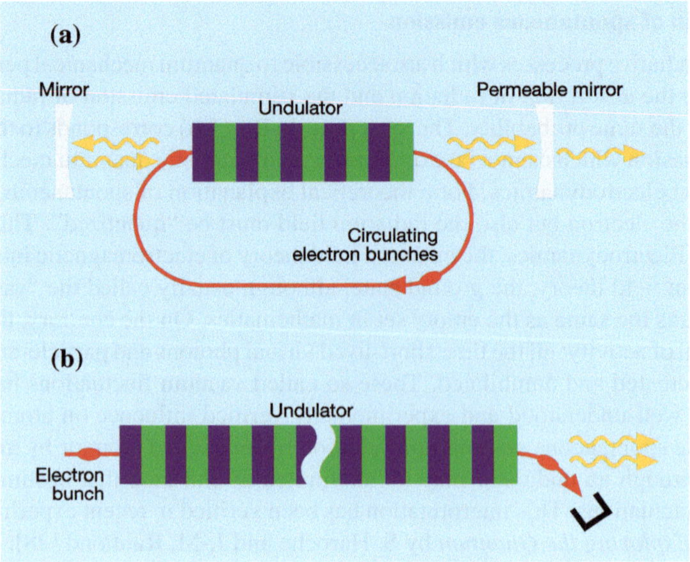

Fig. 1.3 Principle of free-electron laser. For visible or infrared light an optical resonator can be used. A increase in light intensity of a few per cent per passage of a short undulator magnet is sufficient to achieve laser saturation within many round trips. In the ultraviolet and X-ray region one can apply the mechanism of Self-Amplified Spontaneous Emission where a large laser gain is achieved in a single passage through a very long undulator. **a** Low gain FEL. **b** SASE FEL.

The equations of the free-electron laser can be derived from classical relativistic electrodynamics without using the methods of quantum theory. Unlike for optical transitions in atoms, the computation of the power radiated in an undulator or an FEL needs no quantum mechanical matrix elements but can be traced back to the classical Larmor formula for radiation by an accelerated charge. Of course, when the number of photons is of interest or the change in energy and momentum which an electron experiences upon the emission of a photon, the fundamental Planck relation $W = \hbar\omega$ must be used. A genuine quantum theoretical treatment of the FEL [30–32] is only needed under extreme conditions which are not fulfilled in presently existing or planned FEL facilities.

1.3.3 Why Does the FEL Need an Undulator?

We have seen that a relativistic electron beam acts as the energy pump in a free-electron laser. The energy transfer takes place in an undulator magnet, but one may ask why this magnet is needed. Is it conceivable that energy is transferred from the charged particle beam to the light wave when both move on the same straight line? The answer is of course no: in straight-line motion the electric force acting on an electron is always perpendicular to its velocity because electromagnetic waves in vacuum are transverse, and from $-e\,\boldsymbol{E} \cdot \boldsymbol{v} = 0$ follows that no work can be done by the particle. The undulator field forces the electrons on a sinusoidal orbit where the particles acquire a velocity component along the direction of the field, hence $-e\,\boldsymbol{E}\cdot\boldsymbol{v} \neq 0$ and energy transfer will occur.

What about the magnetic force exerted by the light wave? In a high-gain FEL the magnetic field of the light wave may become much larger than the undulator field, and it acts over a large distance since the relativistic particles have a speed v very close to c and move synchronously with the wave. Why does the Lorentz force exerted by the light-wave field not completely spoil the sinusoidal electron orbit in the undulator? The answer is simple: the Lorentz force exerted by the light wave is almost perfectly balanced by the electric force exerted by the wave. To see this we simplify the problem by ignoring the sinusoidal orbit in the undulator. Let the beams move straight along the z direction, i.e. $\boldsymbol{v} = v\,\boldsymbol{e}_z$, and choose a horizontally polarized wave with $\boldsymbol{E} = E\,\boldsymbol{e}_x$, $\boldsymbol{B} = (E/c)\,\boldsymbol{e}_y$. The symbols $\boldsymbol{e}_x$, $\boldsymbol{e}_y$, $\boldsymbol{e}_z$ denote the Cartesian unit vectors. The forces are

$$\boldsymbol{F}_{\rm mag} = -e\,\boldsymbol{v}\times\boldsymbol{B} = +e\,v\,(E/c)\,\boldsymbol{e}_x\,, \quad \boldsymbol{F}_{\rm el} = -e\,\boldsymbol{E} = -e\,E\,\boldsymbol{e}_x\,.$$

Obviously $\boldsymbol{F}_{\rm mag} + \boldsymbol{F}_{\rm el} \to 0$ for $v \to c$. As a consequence of the electrons being at ultra-relativistic energies, any impact of the electromagnetic wave on the trajectory of the electrons can be safely neglected, which will be done in the remainder of this book.

References

1. J.M.J. Madey, Stimulated emission of bremsstrahlung in a periodic magnetic field. J. Appl. Phys. **42**, 1906 (1971)
2. L.R. Elias et al., Observation of stimulated emission of radiation by relativistic electrons in a spatially periodic transverse magnetic field. Phys. Rev. Lett. **36**, 717 (1976)
3. D.A.G. Deacon et al., First operation of a free-electron laser. Phys. Rev. Lett. **38**, 892 (1977)
4. C. Pellegrini, Progress toward a soft X-Ray FEL. Nucl. Instrum. Meth. A **272**, 364 (1988)
5. C. Pellegrini, J. Stöhr, X-ray free-electron lasers—principles, properties and applications. Nucl. Instrum. Meth. A **500**, 33 (2003)
6. C. Pellegrini et al., The SLAC soft X-ray high power FEL. Nucl. Instrum. Meth. A **341**, 326 (1994)
7. P.V. Nickles et al., Short pulse X-Ray laser at 32.6 nm based on transient gain in Ne-like Titanium. Phys. Rev. Lett. **78**, 2748 (1997)
8. P.V. Nickles, K.A. Janulewicz, W. Sandner, X-ray lasers, *Landoldt-Börnstein—Group VIII Advanced Materials and Technologies* (Springer, Heidelberg, 2008), p. 203
9. J.J. Rocca, Table-top soft x-ray lasers. Rev. Scient. Instrum. **70**, 3799 (1999)
10. P. Zeitoun et al., A high-intensity highly coherent soft X-ray femtosecond laser seeded by a high harmonic beam. Nature **431**, 426 (2004)
11. G. Dattoli, A. Renieri, F. Romanelli, Progress in the hamiltonian picture of the free-electron laser. IEEE J. Quantum Electron. **17**, 1371 (1981)
12. R. Bonifacio, F. Casagrande, C. Pellegrini, Hamiltonian model of a free-electron laser. Opt. Comm. **61**, 55 (1987)
13. W.B. Colson, C. Pellegrini, A. Renieri (eds.), Free electron lasers, *Laser Handbook*, vol. 6 (North Holland, Amsterdam, 1990)

14. J.B. Murphy, C. Pellegrini, Introduction to the physics of the free electron laser, *Laser Handbook* (1990), p. 11
15. W.B. Colson, Classical free electron laser theory, *Laser Handbook* (1990), p. 115
16. C.A. Brau, *Free-Electron Lasers* (Academic Press, Boston, 1990)
17. H.P. Freund, T.M. Antonsen, *Principles of Free-Electron Lasers* (Chapman & Hall, London, 1996)
18. P. O'Shea, H.P. Freund, Free-electron lasers: status and applications. Science **292**, 1853 (2001)
19. C. Pellegrini, S. Reiche, The development of X-ray free-electron lasers. IEEE J. Quantum Electron. **10**(6), 1393 (2004)
20. C. Pellegrini, S. Reiche, Free-electron lasers, *The Optics Encyclopedia* (Wiley, Berlin, 2003), p. 1111
21. E.L. Saldin, E.A. Schneidmiller, M.V. Yurkov, *The Physics of Free Electron Lasers* (Springer, Berlin, Heidelberg, 2000)
22. Z. Huang, K.-J. Kim, Review of x-ray free-electron laser theory. Phys. Rev. ST Accel. Beams **10**, 034801 (2007)
23. K.-J. Kim, Z. Huang, R. Lindberg, *Principles of Free Electron Lasers*, US Particle Accelerator School, to be published.
24. J.D. Jackson, *Classical Electrodynamics*, 3rd edn. (John Wiley, New York, 1999)
25. P.J. Duke, *Synchrotron Radiation: Production and Properties* (Oxford University Press, Oxford, 2000)
26. H. Wiedemann, *Particle Accelerator Physics II*, 2nd edn. (Springer, Berlin, 1999)
27. J.A. Clarke, *The Science and Technology of Undulators and Wigglers* (Oxford University Press, Oxford, 2004)
28. S. Haroche, J.-M. Raimond, *Exploring the Quantum* (Oxford University Press, Oxford, 2006)
29. Y.V. Shvyd'ko et al., High-reflectivity high-resolution X-ray crystal optics with diamond. Nat. Phys. **6**(3), 196 (2010)
30. R. Bonifacio, F. Casagrande, Instabilities and quantum initiation in the free-electron laser. Opt. Comm. **50**, 251 (1984)
31. G. Dattoli, A. Renieri, The quantum-mechanical analysis of the free electron laser, *Laser Handbook*. vol. 6 (1990)
32. C.B. Schroeder, C. Pellegrini, P. Chen, Quantum effects in high-gain free-electron lasers. Phys. Rev. E **64**, 056502 (2001)

Chapter 2
Undulator Radiation

2.1 Magnetic Field of a Planar Undulator

The motion of an electron in a planar undulator magnet is shown schematically in Fig. 2.1. The undulator axis is along the direction of the beam (z direction), the magnetic field points in the y direction (vertical). The period λ_u of the magnet arrangement is about 20–30 mm. For simplicity we assume that the horizontal width of the pole shoes is larger than λ_u, then one can neglect the x dependence of the field in the vicinity of the tightly collimated electron beam. The curl of the magnetic field vanishes inside the vacuum chamber of the electron beam, hence the field can be written as the gradient of a scalar magnetic potential obeying the Laplace equation

$$\boldsymbol{B} = -\nabla \Phi_{\text{mag}}\,, \quad \nabla^2 \Phi_{\text{mag}} = 0\,.$$

The field on the axis is approximately harmonic. Making the Ansatz

$$\Phi_{\text{mag}}(y, z) = f(y)\sin(k_u z) \quad \Rightarrow \quad \frac{d^2 f}{dy^2} - k_u^2 f = 0 \ \text{ with } \ k_u = \frac{2\pi}{\lambda_u}$$

we obtain the general solution

$$f(y) = c_1 \sinh(k_u y) + c_2 \cosh(k_u y)\,.$$

The vertical field is

$$B_y(y, z) = -\frac{\partial \Phi_{\text{mag}}}{\partial y} = -k_u\,\left[c_1 \cosh(k_u y) + c_2 \sinh(k_u y)\right]\,\sin(k_u z)\,.$$

B_y has to be symmetric with respect to the plane $y = 0$, hence $c_2 = 0$. We set $k_u c_1 = B_0$ and obtain $B_y(0, z) = -B_0 \sin(k_u z)$. So the potential is

P. Schmüser et al., *Free-Electron Lasers in the Ultraviolet and X-Ray Regime*,
Springer Tracts in Modern Physics 258, DOI: 10.1007/978-3-319-04081-3_2,

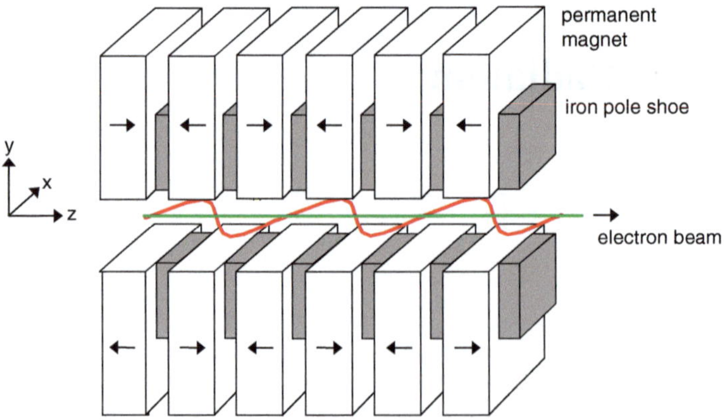

Fig. 2.1 Schematic view of a planar undulator magnet with alternating polarity of the magnetic field and of the sine-like trajectory of the electrons. In the magnet shown here the field is produced by permanent magnets that are placed between iron pole shoes. The distance between two equal poles is called the undulator period λ_u. A typical value is $\lambda_u = 25$ mm.

$$\Phi_{\text{mag}}(x, y, z) = \frac{B_0}{k_u} \sinh(k_u y) \sin(k_u z) \,. \tag{2.1}$$

For $y \neq 0$ the magnetic field possesses a small longitudinal component B_z:

$$\begin{aligned} B_x &= 0 \,, \\ B_y &= -B_0 \cosh(k_u y) \sin(k_u z) \,, \\ B_z &= -B_0 \sinh(k_u y) \cos(k_u z) \,. \end{aligned} \tag{2.2}$$

In the following we restrict ourselves to the symmetry plane $y = 0$ and use the idealized field

$$\boxed{\boldsymbol{B} = -B_0 \sin(k_u z) \, \boldsymbol{e}_y \,,} \tag{2.3}$$

where $\boldsymbol{e}_y$ is the unit vector in y direction.

2.2 Electron Motion in an Undulator

2.2.1 Trajectory in First Order

We call $W = W_{\text{kin}} + m_e c^2 = \gamma m_e c^2$ the total relativistic energy of the electron. The transverse acceleration by the Lorentz force is

$$\gamma m_e \dot{\boldsymbol{v}} = -e \, \boldsymbol{v} \times \boldsymbol{B} \,, \tag{2.4}$$

resulting in two coupled equations

$$\ddot{x} = \frac{e}{\gamma m_e} B_y \dot{z} \,, \qquad \ddot{z} = -\frac{e}{\gamma m_e} B_y \dot{x} \,, \tag{2.5}$$

which are solved iteratively. To obtain the first-order solution we observe that $v_z = \dot{z} \approx v = \beta c =$ const and $v_x \ll v_z$. Then $\ddot{z} \approx 0$ and the solution for $x(t)$ and $z(t)$ is

$$x(t) \approx \frac{e B_0}{\gamma m_e \beta c k_u^2} \sin(k_u \beta c t) \,, \qquad z(t) \approx \beta c t \,, \tag{2.6}$$

if the initial conditions

$$x(0) = 0 \,, \quad \dot{x}(0) = \frac{e B_0}{\gamma m_e k_u}$$

are realized by a suitable beam steering system in front of the undulator[1] (the undulator magnet starts at $z = 0$). The electron travels on the sine-like trajectory

$$x(z) = \frac{K}{\beta \gamma k_u} \sin(k_u z) \,. \tag{2.7}$$

In this equation we have introduced the important dimensionless *undulator parameter*

$$\boxed{K = \frac{e B_0}{m_e c \, k_u} = \frac{e B_0 \lambda_u}{2 \pi \, m_e c} = 0.934 \cdot B_0 \, [\mathrm{T}] \cdot \lambda_u \, [\mathrm{cm}] \,.} \tag{2.8}$$

The transverse velocity is

$$v_x(z) = \frac{K c}{\gamma} \cos(k_u z) \,. \tag{2.9}$$

2.2.2 *Motion in Second Order*

Due to the sinusoidal trajectory the longitudinal component of the velocity is not constant. It is given by

$$v_z = \sqrt{v^2 - v_x^2} = \sqrt{c^2(1 - 1/\gamma^2) - v_x^2} \approx c \left(1 - \frac{1}{2\gamma^2} \left(1 + \gamma^2 v_x^2 / c^2\right)\right) .$$

[1] In practice the initial conditions can be realized by augmenting the undulator with a quarter period upstream of the periodic magnet structure and by displacing the electron orbit at $z = -\lambda_u/4$ by $\Delta x = -K/(\beta \gamma k_u)$ with the help of two dipole magnets. A similar arrangement at the rear end restores the beam orbit downstream of the undulator. For an illustration see Ref. [1].

Inserting for $v_x = \dot{x}(t)$ the first-order solution and using the trigonometric identity $\cos^2\alpha = (1 + \cos 2\alpha)/2$, the longitudinal speed becomes

$$v_z(t) = \left(1 - \frac{1}{2\gamma^2}\left(1 + \frac{K^2}{2}\right)\right)c - \frac{cK^2}{4\gamma^2}\cos(2\omega_u t) \tag{2.10}$$

with $\omega_u = \bar{\beta}c\,k_u \approx c\,k_u$. The average longitudinal speed is

$$\boxed{\bar{v}_z = \left(1 - \frac{1}{2\gamma^2}\left(1 + \frac{K^2}{2}\right)\right)c \equiv \bar{\beta}\,c\,.} \tag{2.11}$$

The particle trajectory in second order is described by the equations

$$x(t) = \frac{K}{\gamma k_u}\sin(\omega_u t)\,, \qquad z(t) = \bar{v}_z t - \frac{K^2}{8\gamma^2\,k_u}\sin(2\omega_u t)\,. \tag{2.12}$$

The motion in a helical undulator is treated in Sects. 4.9 and 10.1.

2.3 Emission of Radiation

The radiation emitted by relativistic electrons in a magnetic field is concentrated in a narrow cone with an opening angle of about $\pm 1/\gamma$. The cone is centered around the instantaneous tangent to the particle trajectory. The direction of the tangent varies along the sinusoidal orbit in the undulator magnet, the maximum angle with respect to the axis being

$$\theta_{\max} \approx \left|\frac{dx}{dz}\right|_{\max} = \frac{K}{\beta\gamma} \approx \frac{K}{\gamma}\,. \tag{2.13}$$

Suppose this directional variation is less than $1/\gamma$. Then the radiation field receives contributions from various sections of the trajectory that overlap in space and interfere with each other. The important consequence is: the radiation spectrum in forward direction is not continuous but nearly monochromatic, more precisely, it is composed of a narrow spectral line at a well-defined frequency and its odd higher harmonics. This is the characteristic feature of undulator radiation. The condition to be satisfied is

$$\theta_{\max} \leq \frac{1}{\gamma} \quad \Rightarrow \quad K \leq 1\,. \tag{2.14}$$

Incidentally, this condition can be a bit relaxed, and K values of $2 - 3$ are still acceptable.

On the other hand, if the maximum angle $\theta_{\max}$ exceeds the radiation cone angle $1/\gamma$ by a large factor, which is the case for $K \gg 1$, one speaks of a wiggler magnet.

Wiggler radiation consists of many densely spaced spectral lines forming a quasi-continuous spectrum which resembles the spectrum of ordinary synchrotron radiation in bending magnets. We will not discuss it any further in this book.

2.3.1 Radiation in a Moving Coordinate System

Consider a coordinate system (x^*, y^*, z^*) moving with the average longitudinal speed of the electrons:

$$\bar{v}_z \equiv \bar{\beta} c \,, \quad \bar{\gamma} = \frac{1}{\sqrt{1-\bar{\beta}^2}} \approx \frac{\gamma}{\sqrt{1+K^2/2}} \quad \text{with } \gamma = \frac{W}{m_e c^2} \,. \tag{2.15}$$

The Lorentz transformation from the moving system to the laboratory system reads

$$\begin{aligned} t^* &= \bar{\gamma}(t - \bar{\beta} z/c) \approx \bar{\gamma}\, t(1-\bar{\beta}^2) = t/\bar{\gamma} \,, \\ x^* &= x = \frac{K}{\gamma k_u} \sin(\omega_u t) \,, \\ z^* &= \bar{\gamma}(z - \bar{\beta} c t) \approx -\frac{K^2}{8\gamma k_u \sqrt{1+K^2/2}} \sin(2\omega_u t) \,. \end{aligned}$$

The electron orbit in the moving system is thus

$$x^*(t^*) = a \sin(\omega^* t^*) \,, \quad z^*(t^*) = -a \frac{K}{8\sqrt{1+K^2/2}} \sin(2\omega^* t^*) \tag{2.16}$$

with the amplitude $a = K/(\gamma k_u)$ and the frequency

$$\omega^* = \bar{\gamma}\, \omega_u = \bar{\gamma}\, c\, k_u \approx \frac{\gamma\, c\, k_u}{\sqrt{1+K^2/2}} \,.$$

Note that $\omega_u t = \omega^* t^*$. The motion is depicted in Fig. 2.2. It is mainly a transverse harmonic oscillation with the frequency $\omega^* = \bar{\gamma}\, \omega_u$. Superimposed is a small longitudinal oscillation with twice that frequency. If we ignore the longitudinal oscillation for the time being, the electron will emit dipole radiation in the moving system with the frequency $\omega^* = \bar{\gamma}\, \omega_u$ and the wavelength $\lambda_u^* = \lambda_u/\bar{\gamma}$.

The radiation power from an accelerated charge is given by the well-known Larmor formula

$$P = \frac{e^2}{6\pi \varepsilon_0 c^3} \dot{\boldsymbol{v}}^2 \,, \tag{2.17}$$

see Ref. [2] or any other textbook on classical electrodynamics. For an oscillating charge, $\dot{\boldsymbol{v}}^2$ must be averaged over one period. The Larmor formula is applicable for

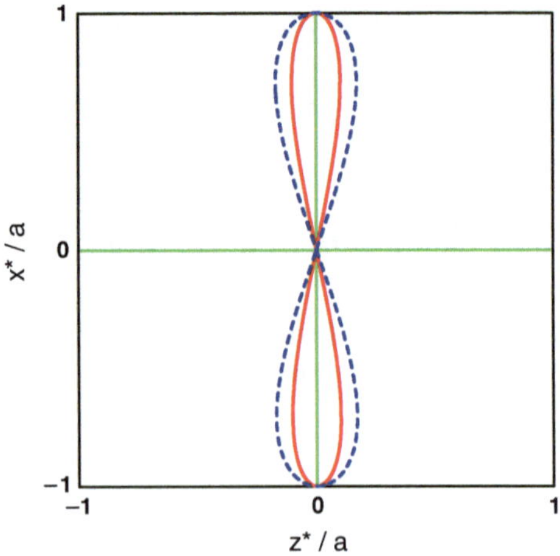

Fig. 2.2 The electron trajectory in the moving coordinate system for an undulator parameter of $K = 1$ (*continuous red curve*) or $K = 5$ (*dashed blue curve*). The *curve* has the shape of the number eight. For $K \gg 1$ the excursion in longitudinal direction is $z^*_{\max}/a = \sqrt{2}/8 = 0.18$. For $K \to 0$ the longitudinal width shrinks to zero.

an oscillating dipole which is either at rest or moving at non-relativistic speeds. This condition is satisfied in the moving coordinate system. Ignoring the longitudinal oscillation, the acceleration has only an x component

$$\dot{v}_x^* = \frac{d^2x^*}{dt^{*2}} = -\frac{K}{\gamma k_u}\,\omega^{*2}\sin(\omega^* t^*) = -\frac{K\gamma c^2 k_u}{1 + K^2/2}\,\sin(\omega^* t^*)\,,$$

and the time-averaged square of the acceleration becomes

$$\left\langle \dot{\boldsymbol{v}}^2 \right\rangle = \frac{K^2\gamma^2 c^4 k_u^2}{(1 + K^2/2)^2}\,\frac{1}{2}\,.$$

The total radiation power in the moving system is thus

$$P^* = \frac{e^2 c\,\gamma^2 K^2 k_u^2}{12\pi\varepsilon_0(1 + K^2/2)^2}\,. \qquad (2.18)$$

2.3.2 Transformation of Radiation into Laboratory System

The radiation characteristics of an oscillating dipole which is either at rest or moving at relativistic speed is depicted in Fig. 2.3. With increasing Lorentz factor γ the radiation becomes more and more concentrated in the forward direction. To compute the light wavelength in the laboratory system as a function of the emission angle θ with respect to the beam axis it is appropriate to apply the Lorentz transformation

$$\hbar\omega^* = \bar{\gamma}\,(W_{\mathrm{ph}} - \bar{\beta}\,c p_{\mathrm{ph}}\cos\theta) = \bar{\gamma}\,\hbar\omega_\ell(1 - \bar{\beta}\cos\theta)$$

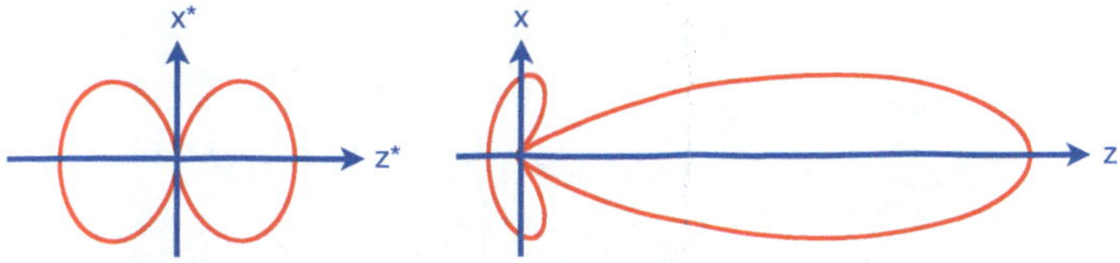

Fig. 2.3 Radiation characteristics in the laboratory system of an oscillating dipole at rest (*left*) or moving horizontally at a speed of $v = 0.9\,c$ (*right*). The dipole oscillates in vertical direction.

which expresses the photon energy $\hbar\omega^*$ in the moving system in terms of the photon energy $W_{ph} = \hbar\omega_\ell$ and the photon momentum $p_{ph} = \hbar\omega_\ell/c$ in the laboratory system. The light frequency in the laboratory system is then

$$\omega_\ell = \frac{\omega^*}{\bar{\gamma}\,(1 - \bar{\beta}\cos\theta)} \quad \Rightarrow \quad \lambda_\ell = \frac{2\pi c}{\omega_\ell} \approx \lambda_u\,(1 - \bar{\beta}\cos\theta)\ .$$

Using $\bar{\beta} = 1 - (1 + K^2/2)/(2\gamma^2)$ and $\cos\theta \approx 1 - \theta^2/2$ (the typical angles are $\theta \le 1/\gamma \ll 1$) we find that the wavelength of undulator radiation near $\theta = 0$ is in good approximation

$$\boxed{\lambda_\ell(\theta) = \frac{\lambda_u}{2\gamma^2}\left(1 + \frac{K^2}{2} + \gamma^2\theta^2\right).} \tag{2.19}$$

The radiation is linearly polarized with the electric vector in the plane of the wavelike electron trajectory. T. Shintake has written a computer code in which the electric field pattern of a relativistic electron moving through the undulator is computed [3]. The field lines are shown in Fig. 2.4. One can clearly see the optical wavefronts and the dependence of the wavelength on the emission angle.

The total radiation power is relativistically invariant [2]. This can be seen as follows. Since we have ignored the longitudinal oscillation, the longitudinal coordinate and the longitudinal momentum of the electron are zero in the moving system

$$z^* = 0\,, \quad p_z^* = 0\ .$$

Then the Lorentz transformations of time and electron energy read

$$t = \bar{\gamma}\,t^*\,, \quad W = \bar{\gamma}\,W^*\,,$$

so the radiation power in the laboratory system becomes

$$P = -\frac{dW}{dt} = -\frac{dW^*}{dt^*} = P^*\ .$$

The undulator radiation power per electron in the laboratory system is therefore

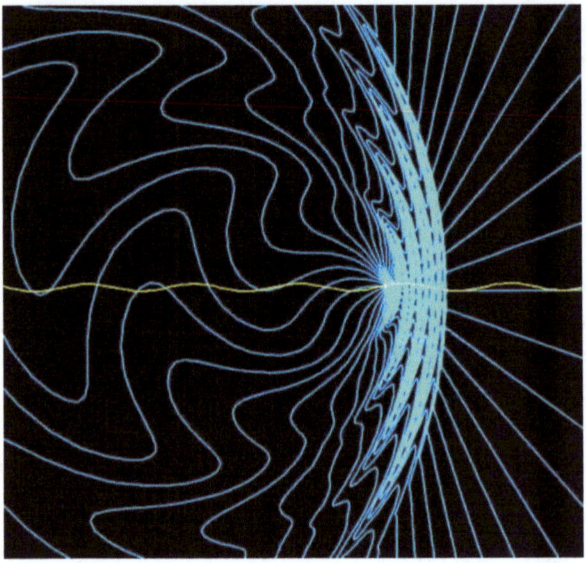

Fig. 2.4 Undulator radiation of an electron with $v = 0.9\,c$. The undulator parameter is $K = 1$. The *wavy curve* indicates the electron trajectory in the undulator. (Courtesy of T. Shintake).

$$P_1 = \frac{e^2 c\,\gamma^2 K^2 k_u^2}{12\pi\varepsilon_0 (1 + K^2/2)^2}\,. \tag{2.20}$$

Since this formula has been derived neglecting the influence of the longitudinal oscillation it describes only the power P_1 contained in the first harmonic. The total power of spontaneous undulator radiation, summed over all harmonics and all angles, is found to be equal to the synchrotron radiation power (1.3) in a bending magnet whose field strength is $B = B_0/\sqrt{2}$.

$$\boxed{P_{\rm rad} = \frac{e^4 \gamma^2 B_0^2}{12\pi\varepsilon_0 c\, m_e^2} = \frac{e^2 c\,\gamma^2 K^2 k_u^2}{12\pi\varepsilon_0}\,.} \tag{2.21}$$

This is easy to understand: the undulator field varies as $B_y(z) = -B_0 \sin(k_u z)$, and hence $\left\langle B_y^2 \right\rangle = B_0^2/2$. Formula (2.21) is valid for any value of the undulator parameter K and therefore also applicable for wiggler radiation. If the undulator parameter is increased much beyond the ideal value of $K = 1$, the power contained in the first harmonic decreases as $1/K^2$.

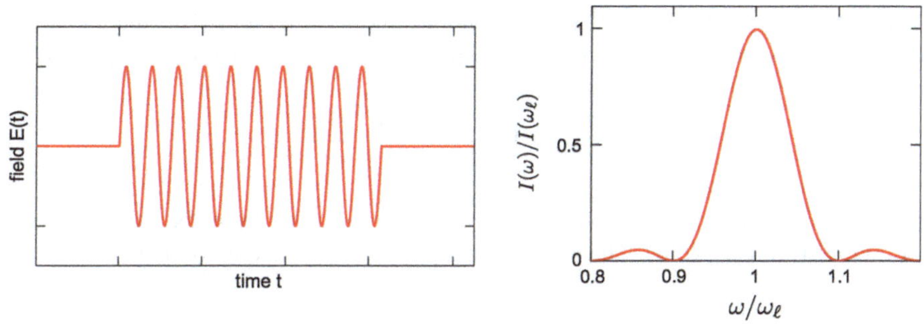

Fig. 2.5 A finite wave train with 10 oscillations and the lineshape function of forward undulator radiation (i.e. the intensity as a function of frequency) in a magnet with $N_u = 10$ periods.

2.4 Lineshape and Spectral Energy of Undulator Radiation

An important property of undulator radiation is that it consists of narrow spectral lines. How wide is such a line? In this section we consider the first harmonic only and look in forward direction. An electron passing through an undulator with N_u periods produces a wave train with N_u oscillations (Fig. 2.5) and a time duration of $T = N_u\lambda_1/c$. The electric field of the light wave is written as

$$E_\ell(t) = \begin{cases} E_0 \exp(-i\,\omega_\ell t) & \text{if } \quad -T/2 < t < T/2\,, \\ 0 & \text{otherwise}\,. \end{cases} \tag{2.22}$$

Due to its finite length, this wave train is not monochromatic but subtends a frequency spectrum which is obtained by Fourier transformation

$$\begin{aligned} A(\omega) &= \int_{-\infty}^{+\infty} E_\ell(t) e^{i\omega t} dt = E_0 \int_{-T/2}^{+T/2} e^{-i(\omega_\ell-\omega)t} dt \\ &= 2E_0 \cdot \frac{\sin((\omega_\ell - \omega)T/2)}{\omega_\ell - \omega}\,. \end{aligned} \tag{2.23}$$

The spectral intensity is

$$I(\omega) \propto |A(\omega)|^2 \propto \left(\frac{\sin\xi}{\xi}\right)^2 \quad \text{with} \quad \xi = \frac{(\omega_\ell - \omega)T}{2} = \pi\, N_u \frac{\omega_\ell - \omega}{\omega_\ell}\,. \tag{2.24}$$

It has a maximum at $\omega = \omega_\ell$ and a characteristic width of

$$\Delta\omega \approx \frac{\omega_\ell}{N_u}\,. \tag{2.25}$$

The lineshape function for a wave train with 10 oscillations is shown in Fig. 2.5.

The angular width of the first harmonic around $\theta = 0$ can be estimated as follows. We know from Eq. (2.19) that the frequency decreases with increasing emission angle θ:

$$\omega_\ell(\theta) = \omega_\ell(0) \cdot \frac{1 + K^2/2}{1 + K^2/2 + \gamma^2\theta^2} \,.$$

The intensity drops to zero when $\delta\omega_\ell = \omega_\ell(0) - \omega_\ell(\theta)$ exceeds the bandwidth following from Eq. (2.24). The root-mean-square (rms) value is found to be [4]

$$\sigma_\theta \approx \frac{1}{\gamma} \cdot \sqrt{\frac{1 + K^2/2}{2N_u}} \approx \frac{1}{\gamma} \cdot \frac{1}{\sqrt{N_u}} \quad \text{for} \quad K \approx 1 \,. \tag{2.26}$$

Obviously, the first harmonic of undulator radiation is far better collimated than synchrotron radiation: the typical opening angle $1/\gamma$ is divided by $\sqrt{N_u} \gg 1$. It is important to realize that this tight collimation of the first harmonic applies only if one requests that the frequency stays within the bandwidth. If one drops the restriction to a narrow spectral line and accepts the entire angular-dependent frequency range as well as the higher harmonics, the cone angle of undulator radiation becomes for $K > 1$, using Eq. (2.13),

$$\theta_{\text{cone}} \approx \frac{K}{\gamma} \,. \tag{2.27}$$

2.5 Higher Harmonics

To understand the physical origin of the higher harmonics of undulator radiation we follow the argumentation in the excellent book *The Science and Technology of Undulators and Wigglers* by J. A. Clarke [5]. In the forward direction ($\theta = 0$) only odd higher harmonics are observed while the off-axis radiation ($\theta > 0$) contains also even harmonics. How can one explain this observation?

Consider a detector with a small aperture centered at $\theta = 0$ which is placed in the far-field at large distance from the undulator. The electrons moving on a sinusoidal orbit with maximum angle of K/γ emit their radiation into a cone of opening angle $1/\gamma$. If the undulator parameter is small, $K \ll 1$, the radiation cone points always toward the detector and therefore the radiation from the entire trajectory is detected. One observes a purely sinusoidal electric field which has only one Fourier component at the fundamental harmonic ω_1, see Fig. 2.6 (top). The situation changes if the undulator parameter is significantly larger than 1, because then the angular excursion of the electron is much larger than the cone angle $1/\gamma$ and the radiation cone sweeps back and forth across the aperture, so the detector receives its light only from short sections of the electron trajectory. The radiation field seen by the detector consists now of narrow pulses of alternating polarity as sketched in the bottom part of Fig. 2.6. The frequency spectrum contains many higher harmonics.

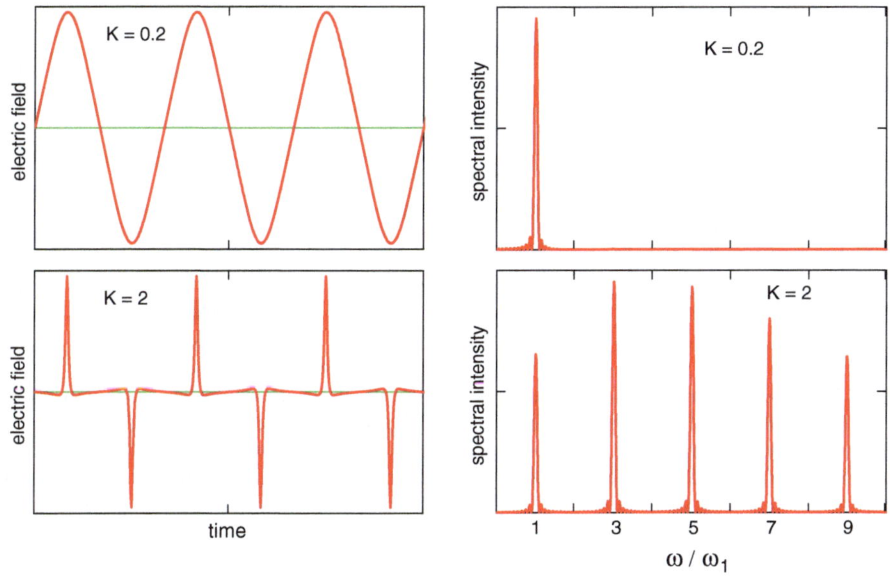

Fig. 2.6 Schematic view of the electric light-wave field seen by a small detector in forward direction and the corresponding frequency spectrum. *Top*: small undulator parameter $K = 0.2$. *Bottom*: fairly large undulator parameter $K = 2$.

Viewed in forward direction, the positive and negative pulses are symmetric in shape and uniformly spaced, and consequently only the odd harmonics occur. When the detector is placed at a finite angle $\theta > 0$, the field pulses are no longer equally spaced and the radiation spectrum contains the even harmonics as well (see [5] for an illustration).

The wavelength of the mth harmonic as a function of the angle θ is

$$\lambda_m(\theta) = \frac{1}{m}\frac{\lambda_u}{2\gamma^2}\left(1 + K^2/2 + \gamma^2\theta^2\right), \quad m = 1, 2, 3, 4, \ldots . \qquad (2.28)$$

In forward direction only the odd harmonics are observed with the wavelengths

$$\boxed{\lambda_m = \frac{1}{m}\frac{\lambda_u}{2\gamma^2}\left(1 + K^2/2\right), \quad m = 1, 3, 5, \ldots ,} \qquad (2.29)$$

so $\lambda_3 = \lambda_1/3$, $\lambda_5 = \lambda_1/5$. We will present an alternative derivation of Eq. (2.29) in Chap. 3.

The spectral energy density per electron of the radiation emitted in forward direction (emission angle $\theta = 0$) is for the mth harmonic [5]

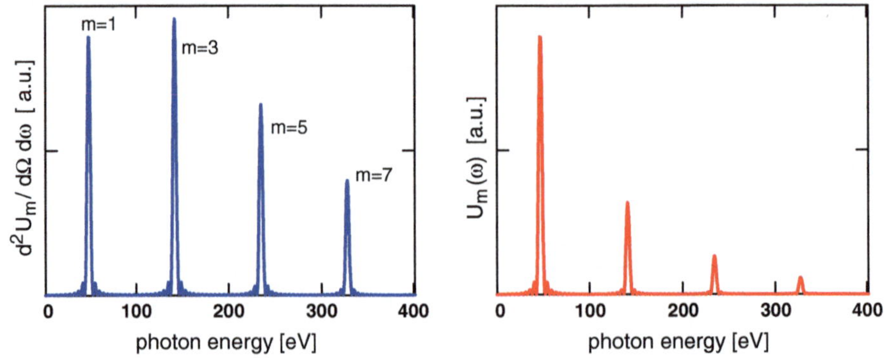

Fig. 2.7 *Left*: Example of a computed photon energy spectrum of undulator radiation for an undulator with 10 periods. Plotted is the differential spectral energy density $d^2U_m/d\Omega d\omega$ at $\theta = 0$. The units are arbitrary. *Right*: The spectral energy $U_m(\omega)$ of the mth harmonic that is emitted into the *solid angle* $\Delta\Omega_m$. The electron Lorentz factor is $\gamma = 1000$, the undulator has the period $\lambda_u = 25$ mm and the parameter $K = 1.5$. Note that the energy ratios U_m/U_1 depend only on the harmonic index m and the undulator parameter K, but neither on γ nor on λ_u.

$$\frac{d^2U_m}{d\Omega d\omega} = \frac{e^2\gamma^2 m^2 K^2}{4\pi\varepsilon_0 c(1+K^2/2)^2} \cdot \frac{\sin^2(\pi N_u(\omega-\omega_m)/\omega_1)}{\sin^2(\pi(\omega-\omega_m)/\omega_1)} \cdot |JJ|^2 \tag{2.30}$$

$$\text{with } JJ = J_n\left(\frac{m\,K^2}{4+2K^2}\right) - J_{n+1}\left(\frac{m\,K^2}{4+2K^2}\right), \quad m = 2n+1\,.$$

Here $\omega_m = m\,\omega_1 \equiv m\,\omega_\ell$ is the (angular) frequency of the mth harmonic. The harmonic index m is related to the index n by $m = 2n + 1$ and takes on the odd integer values $m = 1, 3, 5, \ldots$ for $n = 0, 1, 2, \ldots$. The J_n are the Bessel functions of integer order. In the vicinity of ω_m, the sine function in the denominator of Eq. (2.30) can be replaced by its argument. In this form the equation is presented in Refs. [4, 6].

The absolute bandwidth at $\theta = 0$ is the same for all harmonics

$$\Delta\omega_1 = \Delta\omega_3 = \Delta\omega_5\,, \ldots$$

but the fractional bandwidth shrinks as $1/m$

$$\frac{\Delta\omega_m}{\omega_m} = \frac{1}{mN_u} \tag{2.31}$$

because the wave train comprises now mN_u oscillations in an undulator with N_u periods. The angular width is [4]

$$\sigma_{\theta,m} \approx \frac{1}{\gamma} \cdot \sqrt{\frac{1+K^2/2}{2mN_u}} \approx \frac{1}{\gamma} \cdot \frac{1}{\sqrt{mN_u}} \quad \text{for} \quad K \approx 1\,. \tag{2.32}$$

The corresponding solid angle

$$\Delta\Omega_m = 2\pi\sigma^2_{\theta,m} \approx \frac{2\pi}{\gamma^2}\cdot\frac{1}{m\,N_u}$$

decreases as $1/m$ with increasing harmonic order. Within the solid angle $\Delta\Omega_m$ the angular-dependent frequency shift is less than the bandwidth. Of practical interest is the spectral energy contained in this solid angle:

$$U_m(\omega) = \frac{d^2U_m}{d\Omega d\omega}\,\Delta\Omega_m \qquad m = 1, 3, 5, \ldots\,. \tag{2.33}$$

This spectral energy is shown in Fig. 2.7 for $m = 1, 3, 5, 7$ for a short undulator with ten periods and $K = 1.5$.

The angular dependence of the spectral energy is derived in Ref. [5]. For emission angles $\theta > 0$ the radiation contains all even and odd higher harmonics, as mentioned above.

References

1. K. Wille, *The Physics of Particle Accelerators. An Introduction* (Oxford University Press, Oxford, 2001)
2. J.D. Jackson, *Classical Electrodynamics*, 3rd edn. (Wiley, New York, 1999)
3. T. Shintake, New Real-Time Simulation Technique for Synchrotron and Undulator Radiation. Proceedings of LINAC2002, Geongju, Korea, 2002
4. H. Wiedemann, *Particle Accelerator Physics II*, 2nd edn. (Springer, Berlin, 1999)
5. J.A. Clarke, *The Science and Technology of Undulators and Wigglers* (Oxford University Press, Oxford, 2004)
6. P.J. Duke, *Synchrotron Radiation: Production and Properties* (Oxford University Press, Oxford, 2000)

Chapter 3
Low-Gain FEL Theory

The schematic setup of a low-gain FEL is shown in Fig. 3.1. The main components are

- an electron storage ring or a recirculating linac in which relativistic electron bunches carry out many revolutions,
- a short undulator magnet,
- an optical cavity.

We assume the presence of an initial light wave with wavelength λ_ℓ which may be provided either by an external source such as a conventional laser, or by the spontaneously emitted undulator radiation which is captured in the optical cavity. Following the terminology in laser physics, one speaks of an FEL amplifier if the lasing process is initiated by *seed radiation*, and of an FEL oscillator if the lasing process starts from spontaneous radiation. The bunches make very many turns through the undulator. Upon each passage the light intensity grows by only a few per cent, which is the reason why such a device is called a *low-gain* FEL. The small gain per undulator passage, however, does not prevent the FEL from reaching very high output powers of many Gigawatts if the number of turns is sufficiently large and if the lifetime of the optical eigenmode, being proportional to the quality factor of the optical cavity, is long enough.

3.1 Energy Transfer from Electron Beam to Light Wave

3.1.1 Qualitative Treatment

We consider here the case of an FEL amplifier which is seeded by an external laser. For simplicity, the light wave co-propagating with the relativistic electron beam is described by a plane electromagnetic wave

P. Schmüser et al., *Free-Electron Lasers in the Ultraviolet and X-Ray Regime*,
Springer Tracts in Modern Physics 258, DOI: 10.1007/978-3-319-04081-3_3,

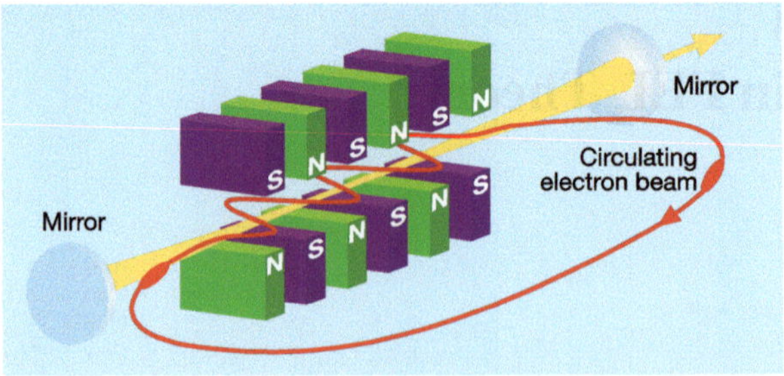

Fig. 3.1 Principle of a low-gain free-electron laser.

$$E_x(z,t) = E_0 \cos(k_\ell z - \omega_\ell t + \psi_0) \quad \text{with} \quad k_\ell = \omega_\ell / c = 2\pi/\lambda_\ell \,. \tag{3.1}$$

(In reality the seed laser will be a pulsed laser to achieve sufficient instantaneous power. Likewise, the FEL light is concentrated in a short pulse that is traveling back and forth between the mirrors of the optical resonator). The electric field of the light wave exerts a force $\boldsymbol{F} = -e\boldsymbol{E}$ on the electron and changes its energy:

$$\frac{dW}{dt} = \boldsymbol{v} \cdot \boldsymbol{F} = -e v_x(t) E_x(t) \,. \tag{3.2}$$

Energy conservation tells us that the light wave gains energy if the electron loses energy, so dW/dt must be negative to get an energy transfer from the electron to the light wave.[1] This happens if the x component of the electron velocity and the electric vector E_x of the light wave point in the same direction. Suppose this condition is fulfilled at some position in the undulator, see Fig. 3.2. Now a problem arises. The light wave, traveling with the speed c along the z axis, will obviously slip forward with respect to the electron whose average speed in z direction is $\bar{v}_z = c\left(1 - (2 + K^2)/(4\gamma^2)\right)$ according to Eq. (2.11). This speed is definitely less than c because the electrons are massive particles and thus slower than light, but more importantly, because they travel on a sinusoidal orbit which is longer than the straight path of the photons.

The question is then: is it possible at all to achieve a sustained energy transfer from the electron beam to the light wave along the entire undulator axis? The answer is: sustained energy transfer is indeed possible but the phase of the light wave has to slip by the right amount, and this proper slippage occurs only for certain well-defined wavelengths. Figure 3.2 illustrates that the transverse velocity v_x and the field E_x remain parallel if the light wave advances by half an optical wavelength $\lambda_\ell/2$ in a half period of the electron trajectory. The difference of the electron and light travel times for a half period of the undulator is

[1] In Chap. 4 the inhomogeneous wave equation will be used to compute the energy exchange between the electron beam and the FEL wave.

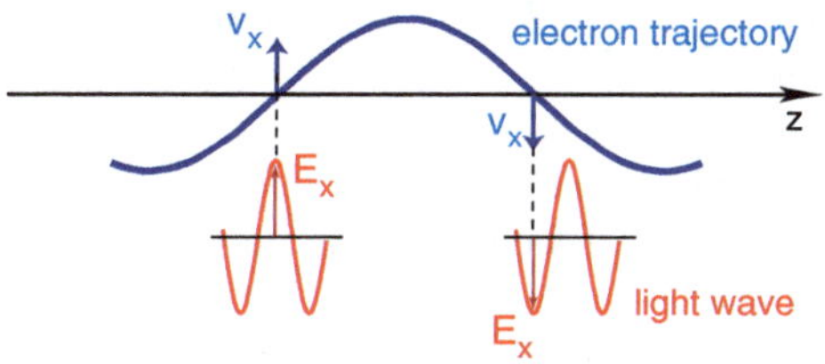

Fig. 3.2 Condition for sustained energy transfer from electron to light wave: the light wave has to advance by $\lambda_\ell/2$ per half period of the electron trajectory.

$$\Delta t = t_{\rm el} - t_{\rm light} = \left[\frac{1}{\bar{v}_z} - \frac{1}{c}\right]\frac{\lambda_u}{2},$$

and the condition for sustained energy transfer reads

$$c\,\Delta t = \lambda_\ell/2\,.$$

Inserting Δt allows us to compute the light wavelength. In good approximation it is given by

$$\lambda_\ell = \frac{\lambda_u}{2\gamma^2}\left(1+\frac{K^2}{2}\right). \tag{3.3}$$

We remark that slippages by $3\lambda_\ell/2,\ 5\lambda_\ell/2\,,\ldots$ are also possible leading to odd higher harmonics ($\lambda_\ell/3\,,\ \lambda_\ell/5\,,\ldots$) of the FEL radiation. Note however that $c\,\Delta t = 2\lambda_\ell/2,\ 4\lambda_\ell/2\,,\ldots$ yields zero net energy transfer from the electron to the light wave: the even harmonics ($\lambda_\ell/2\,,\ \lambda_\ell/4\,,\ldots$) will not be amplified.

3.1.2 Quantitative Treatment

The energy transfer per unit time from an electron to the light wave is described by the equation

$$\frac{dW}{dt} = -e\,v_x(t)E_x(t) = -\frac{e\,c\,K}{\gamma}\cos(k_u z)E_0\cos(k_\ell z - \omega_\ell t + \psi_0) \tag{3.4}$$
$$= -\frac{e\,c\,K\,E_0}{2\gamma}\left[\cos((k_\ell + k_u)z - \omega_\ell t + \psi_0) + \cos((k_\ell - k_u)z - \omega_\ell t + \psi_0)\right].$$

What is the meaning of the phase ψ_0 appearing in the equation? This phase accounts for the fact that the light wave will in general be phase-shifted against the sinusoidal trajectory of an arbitrary electron because the bunch is far longer than the light wavelength. It is convenient to rewrite Eq. (3.4) in the form

$$\frac{dW}{dt} = -\frac{e\,c\,K\,E_0}{2\gamma}\cos\psi - \frac{e\,c\,K\,E_0}{2\gamma}\cos\chi\,, \tag{3.5}$$

where we have introduced two new quantities, ψ and χ. It is customary in FEL physics to call the argument ψ of the first cosine function the *ponderomotive phase*:

$$\psi = (k_\ell + k_u)z - \omega_\ell t + \psi_0\,. \tag{3.6}$$

The position z of the electron is a function of time t according to Eq. (2.12), so we can write the ponderomotive phase as a function of the single variable t:

$$\boxed{\psi(t) = (k_\ell + k_u)z(t) - \omega_\ell t + \psi_0\,.} \tag{3.7}$$

The first term in Eq. (3.5) will provide a continuous energy transfer from the electron to the light wave if $\cos\psi(t)$ is positive and remains constant along the entire undulator (i.e. independent of time), the optimum phase being $\psi = 0$ or $\psi = \pm n\,2\pi$. Now we insert $z(t)$ from Eq. (2.12) but neglect for the time being the longitudinal oscillation, hence we put $z(t) = \bar{v}_z\,t$ and get

$$\psi(t) = (k_\ell + k_u)\bar{v}_z\,t - \omega_\ell t + \psi_0 = \text{const} \quad \Leftrightarrow \quad \frac{d\psi}{dt} = (k_\ell + k_u)\bar{v}_z - k_\ell c = 0\,. \tag{3.8}$$

The condition $\psi = \text{const}$ can only be fulfilled for a certain wavelength. Insertion of $\bar{v}_z$ permits us to compute the light wavelength, which is in good approximation

$$\boxed{\lambda_\ell = \frac{\lambda_u}{2\gamma^2}\left(1 + \frac{K^2}{2}\right).} \tag{3.9}$$

This is an extremely important result: the condition for sustained energy transfer all along the undulator yields the exactly same light wavelength as is observed in undulator radiation at $\theta = 0$. This fact is the reason why spontaneous undulator radiation can serve as seed radiation in an FEL.

Now we look at the second cosine function in Eq. (3.5). Here the argument $\chi(t)$ cannot be kept constant since from the requirement

$$\chi(t) = (k_\ell - k_u)\bar{v}_z t - \omega_\ell t + \psi_0 = \text{const} \tag{3.10}$$

we would deduce

$$k_\ell(1 - \bar{\beta}) = -k_u\,\bar{\beta} \quad \Rightarrow \quad k_\ell < 0\,,$$

which means that the light wave would have to propagate in the negative z direction, contrary to our assumption. Writing ψ as a function of $z = \bar{v}_z\,t$ we can immediately verify that

$$\chi(z) = \psi(z) - 2k_u z\,.$$

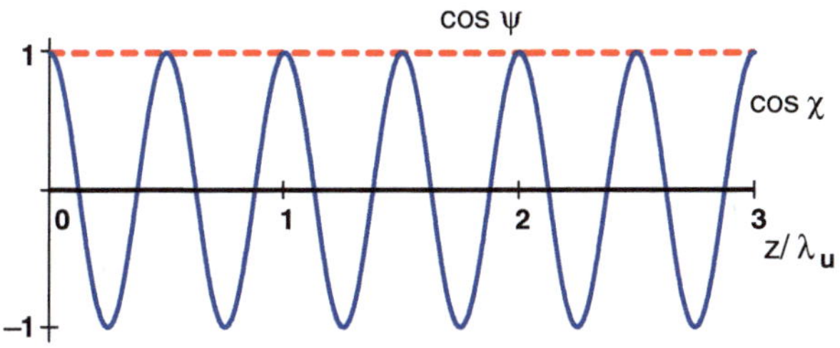

Fig. 3.3 The z dependence of the constant term $\cos\psi$ (*dashed red line*) and the rapidly oscillating term $\cos\chi$ (*blue curve*).

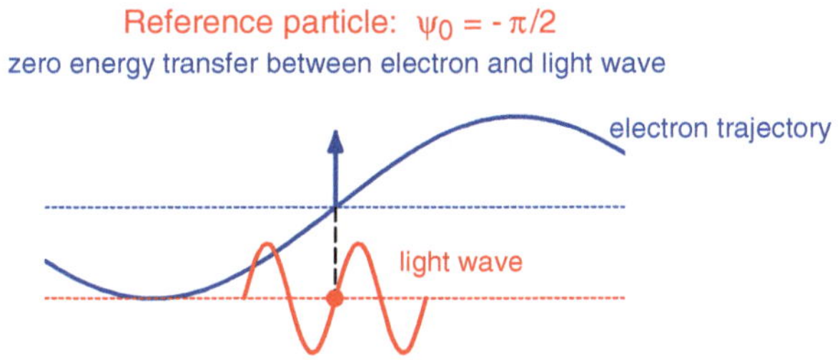

Fig. 3.4 Ponderomotive phase $\psi_0 = -\pi/2$ corresponding to vanishing energy exchange between electron and light wave.

Choosing the Lorentz factor γ of the electrons such that Eq. (3.9) holds, we obtain $\psi(z) = \text{const}$, and then the second cosine function in (3.5) behaves as $\cos(2k_u z)$, i.e. it carries out two oscillations per undulator period and averages to zero. This scenario is depicted in Fig. 3.3. Neglecting the rapidly oscillating second term, Eq. (3.5) reduces to

$$\frac{dW}{dt} = -\frac{ecKE_0}{2\gamma}\cos\psi\,. \tag{3.11}$$

Significance of the ponderomotive phase

To understand the significance of the ponderomotive phase we assume that the light wavelength obeys the basic equation (3.9), from which follows $\psi(z) = \text{const}$. First we choose $\psi_0 = -\pi/2$. In this case both Eq. (3.11) and Fig. 3.4 tell us that there is no energy exchange between electron and light wave.

Choosing an initial phase of $\psi_0 = 0$ (see the left part of Fig. 3.5), we get $\psi(z) = 0$ all along the undulator and thus sustained energy transfer from the electron to the light wave. On the other hand, for $\psi_0 = -\pi$ there will be sustained energy transfer from the light wave to the electron, corresponding to particle acceleration by a light field (right part of Fig. 3.5).

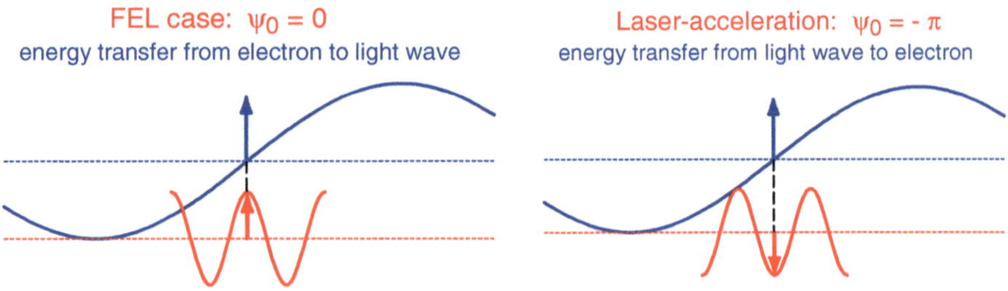

Fig. 3.5 *Left*: Ponderomotive phase $\psi_0 = 0$ for optimum energy transfer from electron to light wave (FEL operation). *Right*: Phase $\psi_0 = -\pi$ for optimum energy transfer from light wave to electron (particle acceleration by a light field).

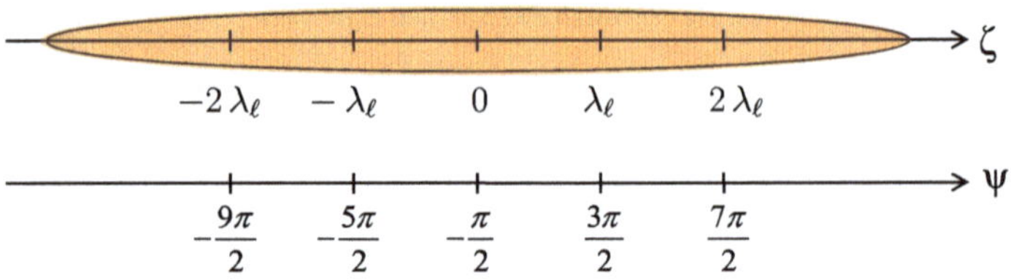

Fig. 3.6 Definition of the internal longitudinal bunch coordinate ζ. The ponderomotive phase is $\psi = 2\pi\,\zeta/\lambda - \pi/2$. The origin $\zeta = 0$ corresponds to an initial phase $\psi_0 = -\pi/2$ in Eqs. (3.4) and (3.6) and vanishing energy exchange between electron and light wave. The origin moves with the average electron speed $\bar{v}_z$ along the z axis of the undulator. Note that the length scale of the coordinate ζ refers to the laboratory system and not to the co-moving coordinate system of the relativistic bunch. Hence there is no relativistic length expansion.

Internal bunch coordinate
The ponderomotive phase ψ has an intuitive interpretation: it can be transformed into a longitudinal coordinate ζ inside the bunch:

$$\boxed{\zeta = \frac{\psi + \pi/2}{2\pi}\,\lambda_\ell\,.} \tag{3.12}$$

This is schematically shown in Fig. 3.6. We define the origin $\zeta = 0$ by the condition that the energy exchange between electron and light wave is zero; this corresponds to an initial phase $\psi_0 = -\pi/2$, see Fig. 3.4. The general electron will be at an initial position $\zeta_0 \neq 0$ in the bunch and it will thus possess an initial phase $\psi_0 \neq -\pi/2$.

3.2 The FEL Pendulum Equations

We treat the low-gain FEL again as a laser amplifier and assume that the lasing process in the undulator is initiated by an incident monochromatic light wave of amplitude E_0 and wavelength λ_ℓ. The *resonance electron energy* $W_r = \gamma_r\, m_e c^2$ is defined by the equations

$$\boxed{\lambda_\ell = \frac{\lambda_u}{2\gamma_r^2}\left(1+\frac{K^2}{2}\right) \quad \Rightarrow \quad \gamma_r = \sqrt{\frac{\lambda_u}{2\lambda_\ell}\left(1+\frac{K^2}{2}\right)}\,.} \tag{3.13}$$

Its meaning is the following: electrons with the energy W_r would emit undulator radiation whose wavelength is identical to the seed wavelength λ_ℓ.

Let now the electron energy W be slightly different from W_r and define the relative energy deviation

$$\boxed{\eta = \frac{W - W_r}{W_r} = \frac{\gamma - \gamma_r}{\gamma_r}\,, \qquad |\eta| \ll 1} \tag{3.14}$$

The Lorentz factor γ and the ponderomotive phase ψ of an electron will both change due to the interaction with the radiation field. In contrast to this, the electric field amplitude E_0 grows so slowly in a low-gain FEL that it can be considered as roughly constant during one passage of the undulator. More precisely, we put $dE_0/dt = 0$ inside the undulator but change the field by the appropriate discrete amount ΔE_0 after each passage. The time derivative of the ponderomotive phase is no longer zero for $\gamma \neq \gamma_r$:

$$\frac{d\psi}{dt} = (k_\ell + k_u)\bar{v}_z - \omega_\ell \approx k_u c - \frac{k_\ell c}{2\gamma^2}\left(1+\frac{K^2}{2}\right),$$

where we have inserted $\bar{v}_z$ from (2.11). According to Eq. (3.13) we can write

$$k_u c = \frac{k_\ell c}{2\gamma_r^2}\left(1+\frac{K^2}{2}\right)$$

and obtain

$$\frac{d\psi}{dt} = \frac{k_\ell c}{2}\left(1+\frac{K^2}{2}\right)\left(\frac{1}{\gamma_r^2} - \frac{1}{\gamma^2}\right).$$

From this follows in good approximation, using Eq. (3.14) and the fact that γ differs very little from γ_r:

$$\boxed{\frac{d\psi}{dt} = 2k_u c\,\eta\,.} \tag{3.15}$$

The time derivative of the relative energy deviation η is according to Eq. (3.11)

$$\boxed{\frac{d\eta}{dt} = -\frac{eE_0 K}{2m_e c\gamma_r^2}\cos\psi\,.} \tag{3.16}$$

The two Eqs. (3.15) and (3.16) are called the FEL pendulum equations. They are of fundamental importance for both low-gain and high-gain FELs.

Analogy with a mathematical pendulum
In order to stress the similarity with the mathematical pendulum it is convenient to introduce a shifted phase variable[2] ϕ by

$$\phi = \psi + \pi/2 \quad \Rightarrow \quad \cos\psi = \sin\phi \, . \tag{3.17}$$

The two coupled first-order differential equations (3.15) and (3.16) read then

$$\frac{d\phi}{dt} = 2k_u c\,\eta \, , \qquad \frac{d\eta}{dt} = -\frac{eE_0 K}{2m_e c\gamma_r^2}\sin\phi \, . \tag{3.18}$$

Combining them we arrive at the second-order pendulum equation of the low-gain FEL

$$\ddot{\phi} + \Omega^2 \sin\phi = 0 \quad \text{with} \quad \Omega^2 = \frac{eE_0 K k_u}{m_e\gamma_r^2} \, , \tag{3.19}$$

which is mathematically equivalent to the second-order differential equation of a mathematical pendulum. We point out that this equation is not valid in the high-gain FEL theory because the time derivative of the electric field has been neglected in the derivation of (3.19).

3.3 Phase Space Representation and FEL Bucket

3.3.1 Phase Space Trajectories

There is a close analogy between the dynamics of a low-gain FEL and the motion of a mathematical pendulum which is treated in Sect. 10.1. The first-order differential equations are

$$\begin{aligned} \textbf{FEL} \quad & \frac{d\phi}{dt} = 2k_u c\cdot\eta \, , & \frac{d\eta}{dt} &= -\frac{eE_0 K}{2m_e c\gamma_r^2}\cdot\sin\phi \, , \\ \textbf{pendulum} \quad & \frac{d\phi}{dt} = \frac{1}{m\ell^2}\cdot L \, , & \frac{dL}{dt} &= -m\,g\,\ell\cdot\sin\phi \, . \end{aligned}$$

Owing to the same mathematical structure of the two sets of coupled equations we can describe the FEL by a Hamiltonian, too. In analogy with Eq. (10.5), the Hamiltonian of the low-gain FEL is given by

$$H(\phi,\eta) = k_u c\,\eta^2 + \frac{eE_0 K}{2m_e c\gamma_r^2}\,(1-\cos\phi) \, . \tag{3.20}$$

---

[2] The shifted phase ϕ is only needed for our comparison between FEL and pendulum. It will not be used in the other chapters.

The Eqs. (3.18) are an immediate consequence of the Hamilton equations

$$\frac{d\phi}{dt} = \frac{\partial H}{\partial \eta} = 2k_u c\,\eta\ , \qquad \frac{d\eta}{dt} = -\frac{\partial H}{\partial \phi} = -\frac{eE_0 K}{2m_e c\gamma_r^2}\sin\phi\ . \tag{3.21}$$

The trajectories in the (ϕ, η) phase space are the curves of a constant Hamiltonian: $H = \text{const}$. The region of bounded motion (periodic oscillation in case of the pendulum) is separated from the region of unbounded motion (rotation of pendulum, see Sect. 10.1) by a curve called the *separatrix*. The equation of the FEL separatrix is in analogy to Eq. (10.7)

$$\eta_{\text{sep}}(\phi) = \pm\sqrt{\frac{eE_0 K}{k_u m_e c^2 \gamma_r^2}}\,\cos(\phi/2)\ . \tag{3.22}$$

The phase space trajectory of an electron in an FEL can be easily constructed by writing the coupled differential equations (3.21) as difference equations and solving these in small time steps. An electron at the "reference position" $\phi = 0$ (corresponding to the bunch center $\zeta = 0$) has zero energy exchange with the light wave as illustrated in Fig. 3.4. If moreover the energy of the reference electron is chosen as $W = W_r = \gamma_r m_e c^2$, hence $\eta = 0$, then the Eqs. (3.21) reveal that the phase space coordinates of this reference electron will be stationary during the motion through the undulator. The point $(\phi, \eta) = (0, 0)$ is therefore a fixpoint in the phase space diagram.

The phase space trajectories of 15 electrons with different initial phases ϕ_0 are shown in Fig. 3.7 for the two cases $\gamma = \gamma_r$ and $\gamma > \gamma_r$. When the electrons are on resonance, $\gamma = \gamma_r$, the net energy transfer is zero since there are as many electrons which supply energy to the light wave as there are electrons which remove energy from the wave. For $\gamma > \gamma_r$, however, the phase space picture clearly shows that there is a positive net energy transfer from the electron beam to the light wave. This will be computed in the next section.

3.3.2 Definition of the FEL Bucket

The particle dynamics in a low-gain FEL has some similarity with the longitudinal dynamics in a proton storage ring. When the proton beam has been accelerated to the design energy, the phase of the radio-frequency (RF) in the accelerating cavities is adjusted such that the bunch receives zero energy gain on average. The energy of a "reference" particle at the bunch center will remain constant. However, individual protons inside the bunch will either be accelerated or decelerated depending on their position relative to the reference particle. The particles carry out *synchrotron oscillations*, these are coupled longitudinal oscillations about the reference position and energy oscillations about the reference energy. The phase space picture looks exactly alike the FEL phase space diagrams in Fig. 3.7.

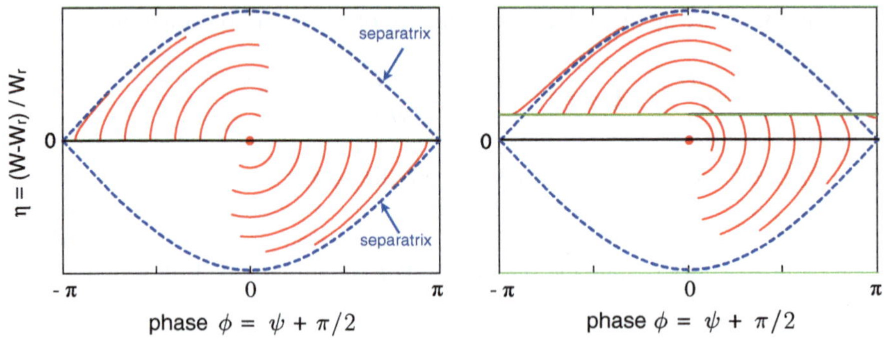

Fig. 3.7 Phase space trajectories of 15 electrons with different initial phases ϕ_0. *Left picture*: Electrons are on resonance, $\gamma = \gamma_r,\ \eta = 0$. The electrons with negative initial phases $-\pi < \phi_0 < 0$ withdraw energy from the light wave while those with positive initial phases $0 < \phi_0 < \pi$ supply energy to the light wave. From the symmetry it is obvious that the net energy transfer from the ensemble of electrons to the light wave is zero for $\gamma = \gamma_r$. The particle at the center $(\phi, \eta) = (0, 0)$ does not move at all, so (0, 0) is a fixpoint. The separatrix (3.22) is drawn as a *dashed curve*. *Right picture*: Electrons are above resonance, $\gamma > \gamma_r,\ \eta > 0$. Now there are more particles losing energy than gaining energy, so the net energy transfer from the electron beam to the light wave is positive (in this example there are seven electrons inside the separatrix that lose energy and five electrons that gain energy). Note that for $\eta > 0$ the fixpoint is not occupied by a particle, and that moreover the first and the last particle are outside the separatrix and carry out an unbounded motion.

The proton bunches are contained in the potential minima of the Hamiltonian, the so-called *RF buckets*. By analogy we call the area enclosed by the separatrix (3.22) the *FEL bucket*. The phase space picture in Fig. 3.7 can be periodically repeated, making the replacement $\phi \rightarrow \phi \pm n\,2\pi$. There are many FEL buckets in a long electron bunch.

Written in terms of the ponderomotive phase $\psi = \phi - \pi/2$ the equation of the separatrix reads

$$\eta_{\text{sep}}(\psi) = \pm\sqrt{\frac{eE_0K}{k_u m_e c^2 \gamma_r^2}}\,\cos\left(\frac{\psi - \psi_b}{2}\right) \tag{3.23}$$

where

$$\psi_b = -\pi/2 \pm n\,2\pi \tag{3.24}$$

is the phase of the bucket center. The energy exchange between electron and light wave vanishes at the bucket center (see also Fig. 3.4).

It is interesting to note that the synchrotron oscillation frequency is quite small. In the 920 GeV proton storage ring HERA at DESY, for example, the particles carried out less than 0.001 longitudinal oscillation per revolution. In a similar manner the phase space motion of the electrons in the FEL is very slow. To see this we rewrite the first pendulum equation in the form

$$\frac{d\psi}{dz} \approx \frac{d\psi}{c\,dt} = \frac{4\pi}{\lambda_u}\eta\,.$$

In a short undulator of 1 m length and with a period of $\lambda_u = 25\,\text{mm}$, the phase advance per passage is only $\Delta\psi = 0.16\,\pi$ for a typical fractional energy deviation $\eta = 0.001$.

3.4 FEL Gain and Madey Theorem

We have seen in the previous section that the energy transfer between the electron beam and the light wave (3.1) depends on the relative energy deviation $\eta = (\gamma - \gamma_r)/\gamma_r$. Figure 3.7 shows quite clearly that the light wave gains energy when η is positive but remains invariant for $\eta = 0$. Likewise, the light wave loses energy when η is negative. Now we look for a quantitative description. The FEL gain function is defined as the relative growth of the light intensity during one passage of the undulator

$$G = \frac{\Delta I_\ell}{I_\ell} \quad \text{with } I_\ell = c\,\frac{\varepsilon_0}{2}\,E_0^2\,.$$

Treating the phase space motion of the electrons by a second-order perturbation approach it can be shown [1] that the gain function[3] is given by the expression

$$G(\omega) = -\frac{\pi\,e^2\widehat{K}^2N_u^3\lambda_u^2\,n_e}{4\varepsilon_0 m_e c^2\gamma_r^3}\cdot\frac{d}{d\xi}\left(\frac{\sin^2\xi}{\xi^2}\right) \quad \text{with} \quad \xi = \xi(\omega) = \pi\,N_u\,\frac{\omega_\ell - \omega}{\omega_\ell}\,. \tag{3.25}$$

Here n_e is the number of electrons per unit volume, N_u is the number of undulator periods, and $\widehat{K}$ is the modified undulator parameter defined in Eq. (3.30) below. The dimensionless variable $\xi = \pi N_u\,(\omega_\ell - \omega)/\omega_\ell$, which was already introduced in Eq. (2.24), is a measure of the frequency deviation from the initial frequency ω_ℓ.

Equation (3.25) is the *Madey theorem* which states that the FEL gain curve is proportional to the negative derivative of the lineshape curve of undulator radiation [2]. We omit here the somewhat cumbersome proof of the Madey theorem because we will demonstrate in Chap. 5 and Sect. 10.2.2 that Eq. (3.25) can be obtained by taking the low-gain limit of the more general high-gain FEL theory.

In Eq. (3.25) the frequency ω is taken as the independent variable. In practice the equation is often applied in a different way. The initial frequency ω_ℓ is fixed by an external seed laser or by the eigenmode of the optical resonator. The electron energy $W = \gamma m_e c^2$, however, may differ from the resonance energy $W_r = \gamma_r m_e c^2$ defined in Eq. (3.13). In this case one is interested in the FEL gain curve as a function of the energy offset. To this end we rewrite the quantity ξ as a function of γ and obtain in good approximation

---

[3] This is the traditional definition of the gain function in FEL theory. In the terminology of electronic amplifiers as well as of standard laser physics the gain should be defined as *gain* $\equiv G + 1$ because unity gain means that the output signal is equal to the input signal.

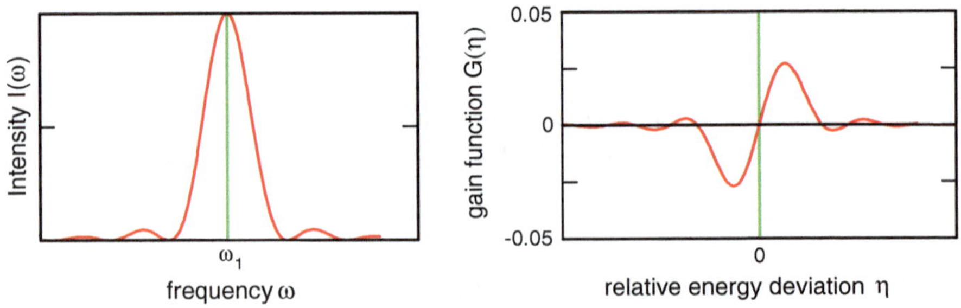

Fig. 3.8 *Left*: the lineshape curve $I(\omega)$ for the first harmonic $\omega_1 \equiv \omega_\ell$ of undulator radiation. *Right*: a typical gain function $G(\eta)$ of the low-gain FEL.

$$\xi = \pi N_u \frac{\omega_\ell - \omega}{\omega_\ell} = -2\pi N_u \frac{\gamma_r - \gamma}{\gamma_r} .$$

The negative sign in front of the last expression is explained in Sect. 5.2.1. The FEL gain as a function of the relative energy deviation η reads then

$$\boxed{G(\eta) = -\frac{\pi e^2 \widehat{K}^2 N_u^3 \lambda_u^2 n_e}{4\varepsilon_0 m_e c^2 \gamma_r^3} \cdot \frac{d}{d\xi}\left(\frac{\sin^2 \xi}{\xi^2}\right) \quad \text{with } \xi = \xi(\eta) = 2\pi N_u \eta .} \tag{3.26}$$

Electrons with positive η enhance the intensity of the light wave, while those with negative η reduce it. An illustration is given in Fig. 3.8.

Qualitatively the Madey theorem is easy to understand. Two conditions must be fulfilled in order to obtain appreciable gain in a free-electron laser driven by seed radiation:

(1) The electron energy W must be close to the resonance energy W_r to enable continuous energy transfer from the electron beam to the light wave all along the undulator. Hence the relative energy deviation $\eta = (W - W_r)/W_r$ must stay within the bandwidth of the lineshape curve of undulator radiation.
(2) From Fig. 3.7 it is obvious the gain function G vanishes for $\eta = 0$, while $G(\eta)$ is positive for $\eta > 0$ and negative for $\eta < 0$.

The negative derivate of the lineshape function satisfies both conditions.

3.5 Higher Harmonics and Modified Undulator Parameter

In the previous sections we have only considered the average longitudinal speed $\bar{v}_z$ of the particles in the computation of the energy exchange between electron and light wave. Now the longitudinal oscillation of $z(t)$ is explicitly taken into account, see Eq. (2.12):

$$z(t) = \bar{v}_z t - \frac{cK^2}{8\gamma^2\omega_u} \sin(2\omega_u t) .$$

We insert this into Eq. (3.4), choosing $\psi_0 = 0$ to simplify the notation:

$$\frac{dW}{dt} = -\frac{ecKE_0}{2\gamma} \left[\cos(k_\ell\, z(t) - \omega_\ell\, t + k_u \bar{v}_z\, t) + \cos(k_\ell\, z(t) - \omega_\ell\, t - k_u \bar{v}_z\, t)\right] .$$

An approximation has already been made here: $k_u z(t) \approx k_u \bar{v}_z\, t$, which is justified since $k_u \ll k_\ell$ and the amplitude of the oscillating term in $z(t)$ is small. At this place it is convenient to write the cosine functions as the real part of the complex exponential function of the form

$$\exp\left[ik_\ell(\bar{\beta} - 1)c\, t \pm ik_u \bar{v}_z\, t\right] \cdot \exp\left[-i\, \frac{K^2 k_\ell}{8\gamma^2 k_u} \sin(2\omega_u t)\right] .$$

The second exponential can be expanded into a Fourier-Bessel series [3]

$$\exp(iY \sin \Phi) = \sum_{n=-\infty}^{\infty} J_n(Y) \exp(in\Phi) \tag{3.27}$$

$$\text{with} \quad Y = -\frac{K^2 k_\ell}{8\gamma^2 k_u} , \qquad \Phi = 2\omega_u t = 2\bar{v}_z k_u t .$$

The two cosine terms in Eq. (3.4) yield now

$$\left[\sum_{n=-\infty}^{\infty} J_n(Y) \exp(i[2n+1]k_u \bar{v}_z t) + \sum_{n'=-\infty}^{\infty} J_{n'}(Y) \exp(i[2n'-1]k_u \bar{v}_z t)\right] \times \exp(ik_\ell[\bar{\beta} - 1]c\, t) .$$

In the second sum we make the replacement $n' \rightarrow (n+1)$. Then the two sums can be combined into a single sum. Taking the real part, the energy change of the electron becomes

$$\frac{dW}{dt} = -\frac{ecKE_0}{2\gamma} \sum_n \left[J_n(Y) + J_{n+1}(Y)\right] \cos\left[(k_\ell + (2n+1)k_u)\bar{v}_z t - k_\ell c\, t\right] .$$

The condition for continuous energy transfer from the electron to the light wave is for the term with index n

$$(k_\ell + [2n+1]k_u)\bar{v}_z - k_\ell c = 0 \quad \Rightarrow \quad k_u = \frac{1}{2n+1} \cdot \frac{k_\ell}{2\gamma^2} \left(1 + \frac{K^2}{2}\right) .$$

Since the light wavelength must be positive only the non-negative integers $n = 0, 1, 2, \ldots$ are allowed. Therefore the FEL wavelengths of the harmonics $m = 2n+1$ are given by the expression

$$\boxed{\lambda_m = \frac{1}{m} \cdot \frac{\lambda_u}{2\gamma^2}\left(1 + \frac{K^2}{2}\right) \quad m = 1,\, 3,\, 5, \ldots\,.} \tag{3.28}$$

We see that only the odd higher harmonics are present. Note this equation is equally valid for undulator radiation in forward direction, see Eq. (2.29).

The energy transfer from electron to light wave is described by the equation

$$\frac{dW}{dt} = -\frac{ecKE_0}{2\gamma}\sum_{n=0}^{\infty}\left[J_n(Y_n) + J_{n+1}(Y_n)\right]\cos\left[(k_\ell + (2n+1)k_u)\bar{v}_z t - k_\ell c\, t\right]$$
$$\text{with} \quad Y_n = -\frac{(2n+1)K^2}{4 + 2K^2}\,. \tag{3.29}$$

The oscillatory term in the longitudinal velocity of the electrons leads not only to the generation of odd higher harmonics but has also an influence on the fundamental harmonic $m = 1$. From Eq. (3.29) follows that the coupling between the charged particle and the electromagnetic wave is changed by the factor $[J_0(Y_0) + J_1(Y_0)]$ if the longitudinal oscillation is taken into consideration. We can absorb this correction factor into a modified undulator parameter[4]

$$\boxed{\widehat{K} \equiv K \cdot \left[J_0\left(\frac{K^2}{4+2K^2}\right) - J_1\left(\frac{K^2}{4+2K^2}\right)\right].} \tag{3.30}$$

Here we have used that J_0 is an even function and J_1 is an odd function. For $K = 1$ the modified undulator parameter is $\widehat{K} = 0.91$.

References

1. K. Wille, *The Physics of Particle Accelerators. An Introduction* (Oxford University Press, Oxford, 2001)
2. J.M.J. Madey, Relationship between mean radiated energy, mean squared radiated energy and spontaneous power spectrum in a power series expansion of the equation of motion in a free electron laser. Nuovo Cimento **50B**, 64 (1979)
3. M. Abramowitz, I.A. Stegun (eds.), *Handbook of Mathematical Functions* (Dover Publications, New York, 1965)

[4] In the FEL literature the modified undulator parameter is often written in the form $K \cdot JJ$ or $K \cdot A_{JJ}$.

Chapter 4
One-Dimensional Theory of the High-Gain FEL

4.1 General Principles of High-Power FELs

There are basically two ways to build a high-power FEL. In the infrared and visible range an optical resonator is utilized, and the FEL radiation pulse passes a short undulator magnet very many times in close spatial and temporal overlap with an electron bunch. During each passage the light intensity grows by just a few per cent but after very many round trips, the overall amplification may become very large. A striking example is the infrared FEL at the Thomas Jefferson Laboratory in Newport News, Virginia, USA where an average FEL beam power of more than 10 kW has been achieved at wavelengths of 6–1.6 μm [1]. It is even possible to reach laser saturation in an FEL equipped with an optical cavity, but one has to realize that the initially uniform particle distribution inside the bunch gradually acquires the microbunch structure discussed below when the light intensity increases more and more. However, when such a bunch traverses the bending magnets of a storage ring or a recirculating linac, the microbunches will be washed out, so a recirculated bunch entering the undulator will not possess a fine structure with the periodicity of the optical wavelength. This bunch may have an increased energy spread owing to the preceding interaction with the light wave.

In the extreme ultraviolet (EUV) and X-ray region an optical resonator has not been realized up to date due to the lack of suitable mirrors.[1] Therefore, the light amplification must be achieved in a single pass through a very long undulator magnet. Here the low-gain FEL theory is obviously inadequate and one has to admit that the amplitude of the light wave grows considerably during the motion through the undulator. The growth is intimately related to the evolution of a microbunch structure at the scale of the light wavelength. This mechanism and the resulting exponential growth and saturation regimes are inaccessible to the low-gain FEL theory presented in Chap. 3.

---

[1] Recently considerable progress has been achieved with crystal optics utilizing Bragg reflection in diamond [2].

P. Schmüser et al., *Free-Electron Lasers in the Ultraviolet and X-Ray Regime*, Springer Tracts in Modern Physics 258, DOI: 10.1007/978-3-319-04081-3_4,

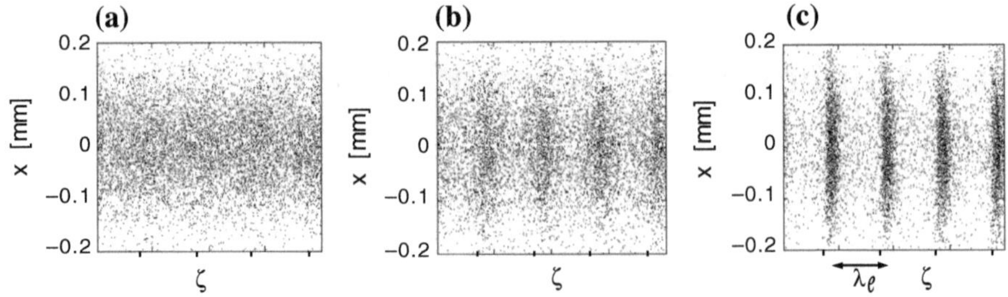

Fig. 4.1 Numerical simulation of microbunching. The particles are plotted as dots in a (x, ζ) plane, where x is the horizontal displacement from the undulator axis and ζ is the longitudinal internal bunch coordinate. **a** Initial uniform distribution, **b** beginning of microbunching, **c** fully developed microbunches with a periodicity of the light wavelength λ_ℓ. (Courtesy of S. Reiche).

4.2 Microbunching

One essential advantage of FEL radiation as compared to undulator radiation is its much higher intensity because a large number of electrons radiate coherently. The intensity of the radiation field grows quadratically with the number of coherently acting particles: $I_N = N^2 I_1$. If it were possible to concentrate all electrons of a bunch into a region far smaller than the light wavelength then these N particles would radiate like a point-like "macroparticle" with charge $Q = -Ne$. The problem is of course that this concentration of some 10^9 electrons into a tiny volume is totally unfeasible, even the shortest conceivable particle bunches are much longer than the wavelength of an X-ray FEL. The way out of this dilemma is given by the process of *microbunching* which is based on the following principle: electrons losing energy to the light wave travel on a sinusoidal trajectory of larger amplitude than electrons gaining energy from the light wave, compare Eq. (2.12). The result is a modulation of the longitudinal velocity which eventually leads to a concentration of the electrons in slices which are shorter than the light wavelength λ_ℓ. These *microbunches* are close to the positions where maximum energy transfer to the light wave can happen according to Fig. 3.2 (we will prove this statement in Sect. 5.6.2).

A numerical simulation of the microbunching process is shown in Fig. 4.1. The particles within a microbunch radiate like a single particle of high charge. The resulting strong radiation field enhances the microbunching even further and leads to an exponential growth of the radiation power. Experimental data at a wavelength of 98 nm are shown in Fig. 4.2. They agree very well with the theoretical prediction.

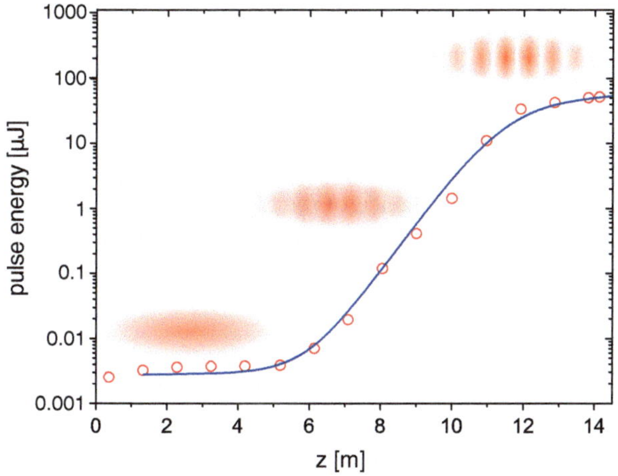

Fig. 4.2 The exponential growth of the FEL pulse energy as a function of the length z traveled in the undulator. The data (*open circles*) were obtained at the SASE FEL of the TESLA Test Facility [3], the electron energy was 245 MeV. The progressing microbunching is indicated schematically. Laser saturation sets in for $z \geq 12$ m. Here the microbunches are fully developed and no further increase in laser power can be expected.

4.3 Basic Elements of the One-Dimensional FEL Theory

In this chapter the high-gain FEL amplifier is studied. The lasing process is initiated by seed radiation of wavelength λ_ℓ and the electron beam energy is chosen to obey the resonance condition (3.13). We restrict ourselves to the one-dimensional FEL theory where a dependency of the bunch charge density and the electromagnetic fields on the transverse coordinates x, y is neglected. This is justified if the electron beam possesses a homogeneous charge density and if its radius r_b is sufficiently large. A lower limit for r_b will be derived in Sect. 6.2. The electron bunches are treated as being very long, the effects occurring at the head or tail of a bunch are ignored. Betatron oscillations and diffraction of the light wave are disregarded as well. The influence of these effects will be investigated in Chap. 6. The full three-dimensional treatment of the FEL is quite complicated and cannot be carried through by analytical methods.

We use complex notation to simplify the mathematics and designate complex quantities with a tilde. For example, the electric field of the light wave inside the undulator is written in the form

$$\boxed{\tilde{E}_x(z,t) = \tilde{E}_x(z)\exp[i(k_\ell z - \omega_\ell t)]} \tag{4.1}$$

with a complex amplitude function $\tilde{E}_x(z)$. The actual field is obtained by taking the real part of this equation

$$E_x(z,t) = \Re\{\tilde{E}_x(z)\exp[i(k_\ell z - \omega_\ell t)]\}\,. \tag{4.2}$$

The analytic description of a high-gain FEL amounts to a self-consistent treatment of

- the coupled pendulum Eqs. (3.15) and (3.16), describing the phase-space motion of the particles under the influence of the electric field of the light wave,
- the inhomogeneous wave equation for the electric field of the light wave,
- the evolution of a microbunch structure coupled with longitudinal space charge forces.

Initially the charge density is distributed almost uniformly along the bunch. During the passage through the undulator, the interaction with the periodic light wave will gradually produce a periodic density modulation. In analogy with the treatment of longitudinal instabilities in circular accelerators [4], we anticipate the microbunching effect by assuming that the initial uniform charge distribution possesses already a small modulation which is periodic in the internal bunch coordinate ζ with the period given by the light wavelength λ_ℓ. From Eq. (3.12) follows then a periodicity in the ponderomotive phase variable ψ with the period 2π. Hence we express the electric charge density in the form

$$\tilde{\rho}(\psi,z) = \rho_0 + \tilde{\rho}_1(z)e^{i\psi}\,. \tag{4.3}$$

The real charge density is of course given by

$$\rho(\psi,z) = \rho_0 + \Re\left(\tilde{\rho}_1(z)e^{i\psi}\right).$$

The complex amplitude $\tilde{\rho}_1 = \tilde{\rho}_1(z)$ grows while the bunch moves through the undulator. From $j_z = v_z\rho$ follows that the current density acquires a similar modulation:

$$\tilde{j}_z(\psi,z) = j_0 + \tilde{j}_1(z)\,e^{i\psi} = j_0 + \tilde{j}_1(z)\,\exp[i(k_\ell + k_u)z - i\omega_\ell t]\,. \tag{4.4}$$

In the high-gain FEL theory we are dealing with many electrons simultaneously which have different positions in the bunch. It is advantageous to take the incident light wave as the reference and put $\psi_0 = 0$ in Eq. (3.6), so the definition of the ponderomotive phase is now

$$\psi = (k_\ell + k_u)z - \omega_\ell\,t\,. \tag{4.5}$$

The initial positions ζ_n of the electrons inside the bunch are specified by assigning a start phase ψ_n to each particle (see Sects. 4.6 and 5.6).

In this chapter we ignore the oscillatory part in the longitudinal velocity and put

$$\boxed{z(t) = \bar{v}_z t = \bar{\beta}c\,t\,, \quad \bar{\beta} = 1 - \frac{1}{2\gamma^2}\left(1 + \frac{K^2}{2}\right).} \tag{4.6}$$

Higher harmonics are therefore not considered in our treatment of the high-gain FEL, but the modified undulator parameter $\widehat{K}$ from Eq. (3.30) is used wherever appropriate, see Sect. 4.5. Using Eqs. (3.9) and (4.6) one gets in good approximation

$$k_\ell + k_u = k_\ell \left[1 + \frac{1}{2\gamma^2}\left(1 + \frac{K^2}{2}\right)\right] \approx \frac{k_\ell}{\bar{\beta}} .$$

Hence the current density can be written as

$$\tilde{j}_z(\psi, z) = j_0 + \tilde{j}_1(z) \exp\left[i k'(z - \bar{v}_z t)\right] \quad \text{with } k' = k_\ell/\bar{\beta} . \tag{4.7}$$

From this equation it is evident that the current modulation moves along the z direction with the same speed $\bar{v}_z = \bar{\beta} c$ as the electron bunch, except for a very slight slippage which is due to the phase evolution of $\tilde{j}_1(z)$. This will be studied in Sect. 5.6.2.

The periodic charge density modulation leads to repulsive space charge forces with the same periodicity. These longitudinal forces counteract the microbunching effect and tend to smear out any structure in the longitudinal charge distribution. Obviously, we have to consider two opposing effects in describing the evolution of the microbunch structure.

The question arises whether it is really justified to assume the existence of an initial periodic density modulation. We can convince ourselves that this is indeed the case. The argument is as follows. The electrons in a bunch entering the undulator are in general randomly distributed along the bunch axis. In Sect. 10.5.1 we show that the random longitudinal distribution has a non-vanishing Fourier component at the wavelength λ_ℓ of the seed radiation. This Fourier component will be strongly amplified in the FEL gain process. Of course one might as well make a Fourier expansion with another wavelength as the period. It will turn out, however, that the FEL gain is large only in a narrow interval around the seed wavelength (see Sect. 5.2), while the gain becomes negligible if λ is very different from λ_ℓ. Hence the Fourier components at other wavelengths will in general not be amplified but retain their small initial values.

Considerable care is needed in applying the Maxwell equations in the 1D FEL theory because these equations are intrinsically three-dimensional. For example, the first Maxwell equation in its integral form states that the flux of the electric field through a closed surface is given by the enclosed electric charge:

$$\oint \boldsymbol{E} \cdot d\boldsymbol{S} = Q/\varepsilon_0 .$$

An interesting consequence is that the internal longitudinal Coulomb force in highly relativistic bunches can be neglected if the bunch length is much larger than its radius (an intuitive explanation is that all particles have velocities very close to c and do not "see" each other). Mathematically we demonstrate this as follows. In the rest frame of the electrons the bunch length appears stretched by the Lorentz factor $\gamma \gg 1$ and is thus very much larger than the bunch radius which remains invariant: $L_b^* = \gamma L_b \gg r_b^* = r_b$. Consequently, the flux of the electric field through a cylindrical surface enclosing the bunch is predominantly in radial direction, the flux through the end faces of the cylinder is almost negligible. Therefore the electric field in the rest frame of the bunch is almost purely radial, and it preserves this feature after Lorentz back-transformation into the laboratory system.

Remark: In contrast to this result, a naive application of the first Maxwell equation in its differential form would lead to an erroneous prediction. What is the 1D approximation of $\mathrm{div}\boldsymbol{E} = \partial E_x/\partial x + \partial E_y/\partial y + \partial E_z/\partial z = \rho/\varepsilon_0$? Since the field is supposed to depend only on z but not on x and y, one might feel inclined to choose the equation equation $dE_z/dz = \rho/\varepsilon_0$. In case of an infinitely long bunch with a constant charge density ρ_0, this equation yields a totally wrong result: it predicts the existence of a longitudinal electric field $E_z(z) = \rho z/\varepsilon_0$ that rises linearly with z and may become very large. On the other hand, for a bunch with a microstructure at the scale of the optical wavelength, described by a charge density $\rho(z) = \rho_0 + \rho_1(z)$, the equation equation $dE_z/dz = \rho_1(z)/\varepsilon_0$ is well suited to compute the longitudinal space charge field, and this will be done below.

In the 1D FEL theory described in this chapter the bunches are assumed to be very long. Then the longitudinal electric field can be neglected as long as the charge distribution remains homogeneous. However, as soon as the periodic charge density modulation sets in at the very tiny scale of the light wavelength, which in the co-moving coordinate frame corresponds to a periodic modulation at the scale of $\lambda_u^* = \lambda_u/\bar{\gamma} \ll L_b^*$, a periodic longitudinal Coulomb field will become significant.

4.4 Electromagnetic Fields

4.4.1 Radiation Field

The wave equation for the electric field $\boldsymbol{E}$ of the light wave reads

$$\left[\nabla^2 - \frac{1}{c^2}\frac{\partial^2}{\partial t^2}\right]\boldsymbol{E} = \mu_0\frac{\partial \boldsymbol{j}}{\partial t} + \frac{1}{\varepsilon_0}\nabla\rho \tag{4.8}$$

with the current density $\boldsymbol{j}$ and the electric charge density ρ. In the one-dimensional approximation the equation for the x component becomes in complex notation

$$\left[\frac{\partial^2}{\partial z^2} - \frac{1}{c^2}\frac{\partial^2}{\partial t^2}\right]\tilde{E}_x(z,t) = \mu_0\frac{\partial \tilde{j}_x}{\partial t}, \tag{4.9}$$

where $\tilde{j}_x$ is the x component of the current density resulting from the sinusoidal motion of the electron bunch in the undulator. The derivative $\partial\rho/\partial x$ is not present in the 1D theory.[2]

We assume that the lasing process in the undulator is initiated (*seeded*) by an incident electromagnetic wave $E_x(z,t)$ with horizontal polarization. In the low-gain FEL, an approximate solution of the wave equation is

$$E_x(z,t) = E_0 \cos(k_\ell z - \omega_\ell t + \psi_0)$$

with a constant amplitude E_0 of the light wave. In reality the amplitude grows by a few per cent during one passage of a short undulator. For the high-gain FEL, however, we have to admit that the amplitude of the light grows considerably along the undulator. Hence we assume a solution of the form

$$\tilde{E}_x(z,t) = \tilde{E}_x(z) \exp[i(k_\ell z - \omega_\ell t)] \tag{4.10}$$

with a complex amplitude $\tilde{E}_x(z)$ which will be a function of the path length z in the undulator. The initial phase ψ_0 is put to zero, see the previous section. The phase of the amplitude $\tilde{E}_x(z)$ may vary with z which means that the phase velocity of the FEL light wave may differ slightly from the phase velocity c of an electromagnetic wave. This effect will be studied in Chap. 6. Inserting Eq. (4.10) into the wave equation yields

$$\left[2\,i\,k_\ell \tilde{E}'_x(z) + \tilde{E}''_x(z)\right] \exp[i(k_\ell z - \omega_\ell t)] = \mu_0 \frac{\partial \tilde{j}_x}{\partial t} \,. \tag{4.11}$$

To proceed further we make the *slowly varying amplitude* (SVA) approximation in order to get rid of the second derivative $\tilde{E}''_x$. The amplitude $\tilde{E}_x(z)$ is assumed to be a smooth function of z which varies slowly in the sense that the change within one undulator period λ_u is small. Then the change within one light wavelength is even much smaller. This implies that the first derivative of the field is also small

$$\left|\tilde{E}'_x(z)\right| \lambda_\ell \ll \left|\tilde{E}_x(z)\right| \quad \Rightarrow \quad \left|\tilde{E}'_x(z)\right| \ll k_\ell \left|\tilde{E}_x(z)\right| . \tag{4.12}$$

Of course, the first derivative, characterizing the slope of the function $\tilde{E}_x(z)$, must be retained in order to describe the growth of the FEL power as a function of undulator length. The change of the slope, however, will be extremely small in one undulator period λ_u and practically negligible in one optical wavelength

$$\left|\tilde{E}''_x(z)\right| \lambda_\ell \ll \left|\tilde{E}'_x(z)\right| \quad \Rightarrow \quad \left|\tilde{E}''_x(z)\right| \ll k_\ell \left|\tilde{E}'_x(z)\right| . \tag{4.13}$$

---

[2] The derivative $\partial\rho/\partial x$ plays no role in the 3D theory either, see Sect. 10.4 and Ref. [5], if one considers only the first harmonic. However, this term becomes important for higher harmonics.

Hence one can safely omit the second derivative of the field in Eq. (4.11). The differential equation of the slowly varying amplitude reads in the SVA approximation

$$\frac{d\tilde{E}_x}{dz} = -\frac{i\mu_0}{2k_\ell} \cdot \frac{\partial \tilde{j}_x}{\partial t} \cdot \exp[-i(k_\ell z - \omega_\ell t)] \,. \tag{4.14}$$

Now a connection is made between the longitudinal and transverse components of the current density. This relation is given by the motion of the particles in the undulator. From

$$\tilde{j}_x = \tilde{\rho} v_x \,, \quad \tilde{j}_z = \tilde{\rho} v_z$$

follows

$$\tilde{j}_x = \tilde{j}_z \, v_x / v_z \,.$$

The x component of the electron velocity in a planar undulator is taken from Eq. (2.9). We get

$$\tilde{j}_x = \tilde{j}_z \frac{v_x}{v_z} \approx \tilde{j}_z \frac{v_x}{c} = \tilde{j}_z \frac{K}{\gamma} \cos(k_u z) \,. \tag{4.15}$$

This relation between the currents in x and z direction is put into Eq. (4.14):

$$\frac{d\tilde{E}_x}{dz} = -\frac{i\mu_0 K}{2k_\ell \gamma} \cdot \frac{\partial \tilde{j}_z}{\partial t} \exp[-i(k_\ell z - \omega_\ell t)] \cos(k_u z) \,.$$

The current density

$$\tilde{j}_z(\psi, z) = j_0 + \tilde{j}_1(z) \exp(i\psi) = j_0 + \tilde{j}_1(z) \exp[i(k_\ell z - \omega_\ell t) + ik_u z]$$

has to be partially differentiated with respect to time

$$\left[\frac{\partial \tilde{j}_z}{\partial t}\right]_{z=const} = -i\omega_\ell \tilde{j}_1 \exp[i(k_\ell z - \omega_\ell t) + ik_u z] \,.$$

The derivative of the transverse field becomes

$$\begin{aligned} \frac{d\tilde{E}_x}{dz} &= -\frac{\mu_0 c K}{2\gamma} \tilde{j}_1 \exp[i(k_\ell z - \omega_\ell t) + ik_u z] \exp[-i(k_\ell z - \omega_\ell t)] \frac{e^{ik_u z} + e^{-ik_u z}}{2} \\ &= -\frac{\mu_0 c K}{4\gamma} \tilde{j}_1 \left\{1 + \exp(i2k_u z)\right\} \,. \end{aligned}$$

The phase factor $\exp(i2k_u z)$ carries out two oscillations per undulator period λ_u and averages to zero (see also Fig. 3.3). So within the SVA approximation we can express the derivative of the transverse field in terms of the modulation amplitude of the current density by

$$\frac{d\tilde{E}_x}{dz} = -\frac{\mu_0 c K}{4\gamma_r} \cdot \tilde{j}_1 \,. \tag{4.16}$$

In this equation we have replaced γ by γ_r because a high-gain FEL is always operated close to resonance (see Sect. 5.1).

4.4.2 Space Charge Field

The longitudinal space charge field created by the modulated charge density (4.3) is computed using the Maxwell equation $\nabla \cdot \boldsymbol{E} = \rho/\varepsilon_0$. According to the discussion in Sect. 4.3 the homogeneous part ρ_0 of the charge density has to be disregarded here. The periodic part of the charge density generates a periodic longitudinal field whose derivative is

$$\frac{\partial \tilde{E}_z(z,t)}{\partial z} = \frac{\tilde{\rho}_1(z)}{\varepsilon_0} \exp[i((k_\ell + k_u)z - \omega_\ell t)] \,. \tag{4.17}$$

Writing

$$\tilde{E}_z(z,t) = \tilde{E}_z(z) \exp[i((k_\ell + k_u)z - \omega_\ell t)]$$

the derivative of the field with respect to z is in the SVA approximation

$$\frac{\partial \tilde{E}_z(z,t)}{\partial z} \approx i\,(k_\ell + k_u)\tilde{E}_z(z) \exp[i((k_\ell + k_u)z - \omega_\ell t)]$$

since

$$\left| \frac{d\tilde{E}_z(z)}{dz} \right| \ll (k_\ell + k_u) \left| \tilde{E}_z(z) \right|$$

for a slowly varying amplitude. Comparing with Eq. (4.17) and using $k_u \ll k_\ell$ we obtain for the complex amplitude of the longitudinal electric field

$$\boxed{\tilde{E}_z(z) \approx -\frac{i}{\varepsilon_0 k_\ell} \tilde{\rho}_1(z) \approx -\frac{i\mu_0 c^2}{\omega_\ell} \cdot \tilde{j}_1(z) \,.} \tag{4.18}$$

Combining Eqs. (4.18) and (4.16) allows us to relate the longitudinal field to the derivative of the transverse field

$$\tilde{E}_z(z) = i\,\frac{4\gamma_r c}{\omega_\ell K} \cdot \frac{d\tilde{E}_x}{dz} \,. \tag{4.19}$$

4.5 Corrections Due to the Longitudinal Oscillation

In the previous section we have disregarded the longitudinal oscillation of the electrons during their motion through the undulator. We know from Sect. 3.5 that this oscillation leads to a modification of the energy transfer from the electron to the light wave, see Eq. (3.29). The modified undulator parameter defined in Eq. (3.30)

$$\widehat{K} = K \cdot \left[J_0\left(\frac{K^2}{4+2K^2}\right) - J_1\left(\frac{K^2}{4+2K^2}\right)\right]$$

enters the energy-transfer Eq. (3.16) which has to read

$$\boxed{\frac{d\eta}{dt} = -\frac{eE_0\widehat{K}}{2m_e c\gamma_r^2}\cos\psi\,.} \tag{4.20}$$

Moreover, $\widehat{K}$ appears in the longitudinal component of the electron-beam current density. Therefore in the Eqs. (4.16) and (4.19) the replacement $K \to \widehat{K}$ must be made:

$$\boxed{\frac{d\tilde{E}_x}{dz} = -\frac{\mu_0 c\widehat{K}}{4\gamma_r}\cdot\tilde{j}_1\,.} \tag{4.21}$$

$$\tilde{E}_z(z) = i\,\frac{4\gamma_r c}{\omega_\ell \widehat{K}}\cdot\frac{d\tilde{E}_x}{dz}\,. \tag{4.22}$$

On the other hand, the FEL wavelength Eqs. (3.9) and (3.28) remain unchanged, so they contain the unmodified undulator parameter K.

4.6 The Coupled First-Order Equations

In Sect. 3.3 we have studied how an ensemble of electrons moves in the (ψ, η) phase space due to the interaction with the field of the light wave. The time derivative of an electron's ponderomotive phase (or, equivalently, of its position ζ inside the bunch) is given by the first pendulum Eq. (3.15). Replacing the time t by the longitudinal coordinate z according to $z(t) = \bar{\beta}c\,t$, we get

$$\frac{d\psi}{dz} = 2k_u\eta \qquad (\bar{\beta}\approx 1)\,. \tag{4.23}$$

The change in the relative energy deviation η is described by the second pendulum Eq. (4.20)

$$\frac{d\eta}{dz} = -\frac{eE_0\widehat{K}}{2m_e c^2 \gamma_r^2} \cos\psi \,.$$

In case of the low-gain FEL the electric field amplitude E_0 of the light wave is treated as a constant but now we have to take into account the z-dependence of the field. In our present complex notation we rewrite the second pendulum equation as

$$\left[\frac{d\eta}{dz}\right]_\ell = -\frac{e\widehat{K}}{2m_e c^2 \gamma_r^2} \Re(\tilde{E}_x e^{i\psi}) \,.$$

At this place it is mandatory to take the real part because the relative energy deviation is always a real quantity. The subscript "ℓ" indicates the coupling to the light wave. Onc has to add the energy change due to the interaction between the electron and the space charge (sc) field. The rate of change of the electron energy due to the longitudinal force is

$$\frac{dW}{dt} = \bar{v}_z F_z = -e\,\bar{v}_z \Re(\tilde{E}_z e^{i\psi}) \,.$$

Using $z = \bar{v}_z t$ we find then

$$\left[\frac{d\eta}{dz}\right]_{sc} = -\frac{e}{m_e c^2 \gamma_r} \Re(\tilde{E}_z e^{i\psi}) \,.$$

Combining the two effects yields

$$\frac{d\eta}{dz} = -\frac{e}{m_e c^2 \gamma_r} \Re\left\{\left(\frac{\widehat{K}\tilde{E}_x}{2\gamma_r} + \tilde{E}_z\right) e^{i\psi}\right\} . \tag{4.24}$$

Our goal is now to study the phase space motion of the electrons in a similar manner as in the low-gain case, but to take explicitly into account the growth of the field amplitude $\tilde{E}_x(z)$ of the light wave and the evolution of the space charge field $\tilde{E}_z(z)$. Both are related to the modulation amplitude $\tilde{j}_1(z)$ of the electron beam current density by the Eqs. (4.21) and (4.18):

$$\frac{d\tilde{E}_x}{dz} = -\frac{\mu_0 c\widehat{K}}{4\gamma_r} \cdot \tilde{j}_1(z) \,, \quad \tilde{E}_z(z) = -\frac{i\mu_0 c^2}{\omega_\ell} \cdot \tilde{j}_1(z) \,. \tag{4.25}$$

The obvious task is to compute $\tilde{j}_1$ for a given arrangement of electrons in phase space. To this end we subdivide the electron bunch into longitudinal slices of length λ_ℓ. Since the bunch is much longer than the light wavelength, there will be very many of these slices (in fact infinitely many in the periodic 1D theory where one neglects the beginning and the end of the bunch). Each slice has an area $A_b = \pi r_b^2$, where r_b is the radius of the bunch. Written in terms of the ponderomotive phase these are slices of length 2π. In the slice $0 \le \psi < 2\pi$ we have N electrons with the phases ψ_n $(n = 1 \dots N)$. Treating the electrons as point-like particles the longitudinal distribution can be expressed in the form

$$S(\psi) = \sum_{n=1}^{N} \delta(\psi - \psi_n) \quad \text{with} \;\; \psi,\; \psi_n \in [0, 2\pi]\,. \tag{4.26}$$

In this chapter and in Chap. 5 we restrict ourselves to the special cases that the initial state is characterized either by a perfectly uniform longitudinal distribution of the electrons in the bunch or by a density distribution that is periodic in ψ with the period 2π. To this end we continue the function (4.26) periodically so that it is defined for all $|\psi| < \infty$. We call this the periodic model. The more realistic case of a random longitudinal particle distribution will be investigated in Chap. 7 and Sect. 10.5.1. The uniformity (or periodicity) implies that the function $S(\psi)$ can be expanded in a real Fourier series

$$S(\psi) = \frac{a_0}{2} + \sum_{k=1}^{\infty} \left[a_k \cos(k\,\psi) + b_k \sin(k\,\psi)\right], \tag{4.27}$$

$$a_k = \frac{1}{\pi}\int_0^{2\pi} S(\psi)\cos(k\,\psi)d\psi\,, \quad b_k = \frac{1}{\pi}\int_0^{2\pi} S(\psi)\sin(k\,\psi)d\psi\,.$$

Defining the complex Fourier coefficients $c_k = a_k - ib_k$ one can rewrite $S(\psi)$ in the form

$$S(\psi) = \frac{c_0}{2} + \Re\left\{\sum_{k=1}^{\infty} c_k \exp(i\,k\,\psi)\right\}. \tag{4.28}$$

The complex Fourier coefficients are given by

$$c_k = \frac{1}{\pi}\int_0^{2\pi} S(\psi)\exp(-i\,k\,\psi)d\psi\,. \tag{4.29}$$

In order to find the relation between the Fourier coefficients and the current density we look first at the dc current density which is proportional to the zeroth Fourier coefficient $c_0/2 \equiv a_0/2 = N/(2\pi)$. From $j_0 = -e\,c\,n_e$ and $n_e = N/(A_b\lambda_\ell)$ follows then

$$j_0 = -e\,c\,\frac{2\pi}{A_b\lambda_\ell}\cdot\frac{c_0}{2}\,.$$

Remember that n_e is the number of electrons per unit volume, N is the number of electrons in a slice of area A_b and length λ_ℓ, and j_0 is the dc current density.

The modulation current $\tilde{j}_1$ is proportional to the coefficient of the first harmonic $k = 1$

$$c_1 = \frac{1}{\pi}\int_0^{2\pi}\left(\sum_{n=1}^{N}\delta(\psi-\psi_n)\right)\exp(-i\,\psi)d\psi = \frac{1}{\pi}\sum_{n=1}^{N}\exp(-i\psi_n)\,.$$

By analogy with j_0 we obtain

$$\tilde{j}_1 = -e\,c\,\frac{2\pi}{A_b\lambda_\ell}\cdot c_1 = -e\,c\,\frac{2\pi}{A_b\lambda_\ell}\cdot\frac{1}{\pi}\sum_{n=1}^{N}\exp(-i\psi_n) = j_0\frac{2}{N}\sum_{n=1}^{N}\exp(-i\psi_n)\,. \tag{4.30}$$

Now we have collected all equations that are needed to compute numerically the time evolution of our system. We remind the reader that our time variable is here the path length z in the undulator magnet which is related to the normal time t by $z = \bar{v}_z t$.

The complete set of coupled first-order equations in the periodic model is

$$\frac{d\psi_n}{dz} = 2k_u\eta_n\,,\quad n = 1........N\,, \tag{4.31a}$$

$$\frac{d\eta_n}{dz} = \frac{-e}{m_e c^2\gamma_r}\Re\left\{\left(\frac{\widehat{K}\tilde{E}_x}{2\gamma_r} - \frac{i\mu_0 c^2}{\omega_\ell}\cdot\tilde{j}_1\right)\exp(i\psi_n)\right\}, \tag{4.31b}$$

$$\tilde{j}_1 = j_0\,\frac{2}{N}\sum_{n=1}^{N}\exp(-i\psi_n)\,, \tag{4.31c}$$

$$\frac{d\tilde{E}_x}{dz} = -\frac{\mu_0 c\widehat{K}}{4\gamma_r}\cdot\tilde{j}_1\,. \tag{4.31d}$$

The Eqs. (4.31a, 4.31b, 4.31c, 4.31d) describe the time evolution of the ponderomotive phase ψ_n and the relative energy deviation $\eta_n = (\gamma_n - \gamma_r)/\gamma_r$ of the nth electron ($n = 1.....N$), as well as the time evolution of the modulated current density $\tilde{j}_1$ and the amplitude of the light wave $\tilde{E}_x$. Since N is a large number we are confronted with a true many-body problem for which no analytical solution exists. The set of $2N+2$ coupled differential and algebraic equations can be solved by numerical integration as will be discussed in Sect. 5.4. We point out that the coupled equations in the form (4.31a, 4.31b, 4.31c, 4.31d) are restricted to uniform or periodic initial particle distributions. They are well suited for a simulation of microbunching, gain and saturation in an FEL amplifier that is driven by monochromatic light, but a SASE FEL cannot be handled because here the initial particle distribution is random.

Now we sketch briefly how the periodic model can be generalized. The main changes are:

(a) The initial phase space distribution must be specified for the whole bunch and not just for one slice of length λ_ℓ.

(b) The electron beam current is a function of z, like $\tilde{j}_1(z)$, but depends in addition on the internal bunch coordinate $\zeta = z - \bar{\beta} c t = z - \bar{v}_z t$. The current density is therefore written as

$$\tilde{j}_z = j_0(\zeta) + \hat{j}_1(z, \zeta) \exp(i \psi).$$

(c) To describe variations of the electric field inside the FEL pulse we introduce an internal longitudinal coordinate

$$u = z - c t = \left(1 - \frac{c}{\bar{v}_z}\right) z + \frac{c}{\bar{v}_z} \zeta$$

and make the Ansatz

$$\tilde{E}_x(z, t) = \widehat{E}(z, u) \exp[i(k_\ell z - \omega_\ell t)]$$

in which the complex amplitude function $\widehat{E}(z, u)$ depends on the position z in the undulator, like $\tilde{E}_x(z)$, but also on the internal FEL pulse coordinate $u = z - ct$. The detailed treatment is presented in Sect. 10.2.4 where also a non-periodic form (10.63a, 10.63b, 10.63c, 10.63d) of the coupled equations can be found.

4.7 The Third-Order Equation of the High-Gain FEL

The main physics of the high-gain FEL is contained in the coupled first-order Eqs. (4.31a, 4.31b, 4.31c, 4.31d) presented in the previous section, but unfortunately these equations have no analytical solution. If we make the additional assumption that the periodic density modulation remains small, it is possible to eliminate the quantities ψ_n and η_n, characterizing the particle dynamics in the bunch, and to derive a third-order differential equation containing only the electric field amplitude $\tilde{E}_x(z)$ of the light wave. This equation has the great advantage that it can be solved analytically.

We present here a novel method to derive the third-order equation from the coupled Eqs. (4.31a, 4.31b, 4.31c, 4.31d) by a perturbation approach. The conventional method using a particle distribution function and the Vlasov equation is outlined in Sect. 10.2.1. The z-evolution of the particle phases ψ_n and the energy deviations η_n will be parametrized in terms of two dimensionless complex functions $\tilde{a}(z)$ and $\tilde{b}(z)$ which are assumed to remain small, $|\tilde{a}(z)| \ll 1$, $|\tilde{b}(z)| \ll 1$. Likewise, the modulated current density $\tilde{j}_1(z)$ and the field $\tilde{E}_x(z)$ are assumed to be small. We restrict ourselves to first-order perturbation theory: only those terms are retained which are linear in $\tilde{a}, \tilde{b}, \tilde{j}_1$ and $\tilde{E}_x$ while second-order terms such as $\tilde{a}^2$, $\tilde{a}\tilde{j}_1$ or $\tilde{a}\tilde{E}_x$ are dropped. Another reason for leaving out the second-order terms is that they would lead to higher harmonics in the FEL radiation which we have explicitly excluded from our simplified analysis.

We consider a mono-energetic incident beam whose energy W may be different from the resonance energy W_r, hence $\eta = (W - W_r)/W_r$ may be different from zero. Energy spread is neglected in the present consideration. Furthermore we assume periodic initial phases:

$$\psi_n(0) \equiv \phi_n = 2\pi \frac{n}{N}, \quad n = 0, 1.....(N-1).$$

The number N of particles is usually very large. The following Ansatz is made for the particle phases and energy deviations as functions of z:

$$\psi_n(z) = \phi_n + 2k_u \eta z + \Re\left\{\tilde{a}(z)e^{i\phi_n}\right\}, \tag{4.32}$$

$$\eta_n(z) = \eta + \Re\left\{\tilde{b}(z)e^{i\phi_n}\right\}. \tag{4.33}$$

This Ansatz is substituted in the four coupled Eqs. (4.31a, 4.31b, 4.31c, 4.31d), resulting in four equations for the functions $\tilde{a}(z)$, $\tilde{b}(z)$ and their derivatives. These equations contain $\tilde{E}_x$, $\tilde{E}'_x$, $\tilde{E}''_x$ and $\tilde{E}'''_x$. By eliminating $\tilde{a}(z)$, $\tilde{b}(z)$ and their derivatives we arrive at the third-order equation. The four steps of the computation are:

First step. The expressions (4.32) and (4.33) are substituted in Eq. (4.31a):

$$\begin{aligned}\psi'_n(z) &= 2k_u\eta_n(z)\\ \Rightarrow \quad & 2k_u\eta + \Re\left\{\tilde{a}'(z)e^{i\phi_n}\right\} = 2k_u\eta + \Re\left\{2k_u\tilde{b}(z)e^{i\phi_n}\right\}.\end{aligned}$$

This equality is valid for all phases ϕ_n. For $n = N$ we obtain $e^{i\phi_n} = 1$ and therefore $\Re\left\{\tilde{a}'(z)\right\} = 2k_u\Re\left\{\tilde{b}(z)\right\}$, while for $n = N/4$ we find $e^{i\phi_n} = i$ and $\Im\left\{\tilde{a}'(z)\right\} = 2k_u\Im\left\{\tilde{b}(z)\right\}$. Hence the following relation holds between the complex functions $\tilde{a}'(z)$ and $\tilde{b}(z)$:

$$\tilde{a}'(z) = 2k_u\tilde{b}(z). \tag{4.34}$$

Second step. Now we take Eq. (4.31b)

$$\frac{d\eta_n}{dz} = -\frac{e}{m_e c^2\gamma_r}\Re\left\{\left(\frac{\widehat{K}\tilde{E}_x}{2\gamma_r} - \frac{i\mu_0 c^2}{\omega_\ell}\tilde{j}_1\right)e^{i\psi_n}\right\}$$

and insert again the expressions (4.32) and (4.33):

$$\Re\left\{\tilde{b}'(z)e^{i\phi_n}\right\} = -\frac{e}{m_e c^2\gamma_r}\Re\left\{\left(\frac{\widehat{K}\tilde{E}_x}{2\gamma_r} - \frac{i\mu_0 c^2}{\omega_\ell}\tilde{j}_1\right)e^{i\phi_n}e^{i\,2k_u\eta z}\exp\left(i\,\Re[\tilde{a}(z)e^{i\phi_n}]\right)\right\}.$$

The Taylor expansion of the last exponential yields

$$\exp\left(i\,\Re[\tilde{a}(z)e^{i\phi_n}]\right) \approx 1 + i\,\Re[\tilde{a}(z)e^{i\phi_n}].$$

In our first-order perturbation treatment we drop the bilinear terms that are proportional to $\tilde{E}_x\,\tilde{a}$ and $\tilde{j}_1\,\tilde{a}$. Hence we make the replacements

$$\tilde{E}_x \exp\left(i\,\Re[\tilde{a}(z)e^{i\phi_n}]\right) \to \tilde{E}_x\,, \quad \tilde{j}_1 \exp\left(i\,\Re[\tilde{a}(z)e^{i\phi_n}]\right) \to \tilde{j}_1$$

and obtain

$$\Re\left\{\tilde{b}'(z)e^{i\phi_n}\right\} = -\frac{e}{m_e c^2\gamma_r}\Re\left\{\left(\frac{\widehat{K}\tilde{E}_x}{2\gamma_r} - \frac{i\mu_0 c^2}{\omega_\ell}\tilde{j}_1\right)e^{i\phi_n}e^{i\,2k_u\eta\,z}\right\}.$$

Also this equation holds for all phase factors $e^{i\phi_n}$, and therefore we get

$$\tilde{b}'(z) = -\frac{e}{m_e c^2\gamma_r}\left(\frac{\widehat{K}\tilde{E}_x}{2\gamma_r} - \frac{i\mu_0 c^2}{\omega_\ell}\cdot\tilde{j}_1(z)\right)e^{i\,2k_u\eta\,z}. \tag{4.35}$$

Third step. Equation (4.31c) is rewritten:

$$\begin{aligned}
\tilde{j}_1(z) &= j_0\,\frac{2}{N}\sum_n e^{-i\,\psi_n(z)} = j_0\,e^{-i\,2k_u\eta\,z}\,\frac{2}{N}\sum_n e^{-i\phi_n}\exp\left(-i\,\Re[\tilde{a}(z)e^{i\phi_n}]\right),\\
\tilde{j}_1(z) &\approx j_0\,e^{-i\,2k_u\eta\,z}\,\frac{2}{N}\sum_n e^{-i\phi_n}\left(1 - i\,\Re[\tilde{a}(z)e^{i\phi_n}]\right).
\end{aligned} \tag{4.36}$$

The real part is expressed in the form

$$\Re[\tilde{a}(z)e^{i\phi_n}] = \frac{1}{2}\,(\tilde{a}(z)e^{i\phi_n} + \tilde{a}^*(z)e^{-i\phi_n}).$$

The $\tilde{a}^*$ term in (4.36) vanishes: its coefficient is proportional to $\sum_n e^{-i2\phi_n}$, and this sum of phase factors vanishes identically.[3] Hence

$$\tilde{j}_1(z) = -i\,j_0\,e^{-i\,2k_u\eta\,z}\,\tilde{a}(z)\underbrace{\frac{2}{N}\sum_{n=0}^{N-1}\frac{1}{2}e^{-i\phi_n}e^{+i\phi_n}}_{N/2}.$$

[3] The equality $\sum_n \exp(-i2\phi_n) = \sum_n \exp(-i\,4\pi n/N) = 0$ follows from the sum formula of a geometric series $S_N = \sum_{n=0}^{N-1} q^n = (1-q^N)/(1-q)$. Putting $q = \exp(-i\,4\pi/N)$ we get $q^N = 1$ and $S_N = 0$.

The current density is thus

$$\tilde{j}_1(z) = -i j_0 \, e^{-i\,2k_u \eta z} \, \tilde{a}(z) \, , \tag{4.37}$$

and Eq. (4.31d) reads now

$$\tilde{E}_x'(z) = -\frac{\mu_0 c \widehat{K}}{4\gamma_r} \cdot \tilde{j}_1(z) = i \, \frac{\mu_0 c \widehat{K}}{4\gamma_r} \, j_0 \, e^{-i\,2k_u \eta z} \, \tilde{a}(z) \, . \tag{4.38}$$

Fourth step. Having derived the Eqs. (4.34–4.38) we now eliminate the unknown functions $\tilde{a}(z)$, $\tilde{b}(z)$ and their derivatives as well as the unknown current density modulation $\tilde{j}_1(z)$. Combining (4.34) and (4.35) we get

$$\tilde{a}'' = 2k_u \tilde{b}' = -\frac{2k_u e}{m_e c^2 \gamma_r} \left(\frac{\widehat{K} \tilde{E}_x}{2\gamma_r} - \frac{i \mu_0 c^2}{\omega_\ell} \tilde{j}_1(z) \right) e^{i\,2k_u \eta z} .$$

Now we insert (4.37) and solve for $\tilde{E}_x$:

$$\tilde{E}_x = \frac{2\gamma_r}{\widehat{K}} \, e^{-i\,2k_u \eta z} \left(\frac{\mu_0 c^2 j_0}{\omega_\ell} \, \tilde{a} - \frac{\gamma_r m_e c^2}{2k_u e} \, \tilde{a}'' \right) . \tag{4.39}$$

The derivatives of Eq. (4.38) are

$$\tilde{E}_x'' = i \, \frac{\mu_0 c \widehat{K} j_0}{4\gamma_r} \, e^{-i\,2k_u \eta z} \left(-i\,2k_u \eta \, \tilde{a} + \tilde{a}' \right) , \tag{4.40}$$

$$\tilde{E}_x''' = i \, \frac{\mu_0 c \widehat{K} j_0}{4\gamma_r} \, e^{-i\,2k_u \eta z} \left(-(2k_u \eta)^2 \, \tilde{a} - i\,4k_u \eta \, \tilde{a}' + \tilde{a}'' \right) . \tag{4.41}$$

In order to eliminate $\tilde{a}''$ we multiply Eq. (4.39) with the factor

$$\left(i \, \frac{\mu_0 c \widehat{K} j_0}{4\gamma_r} \right) \cdot \left(\frac{2\gamma_r}{\widehat{K}} \, \frac{\gamma_r m_e c^2}{2k_u e} \right)^{-1} \equiv -i \, \Gamma^3$$

and add it to Eq. (4.41). This yields

$$\tilde{E}_x''' - i \, \Gamma^3 \tilde{E}_x = i \, \frac{\mu_0 c \widehat{K} j_0}{4\gamma_r} \, e^{-i\,2k_u \eta z} \left(\left[\frac{\mu_0 c^2 j_0}{\omega_\ell} \, \frac{2k_u e}{\gamma_r m_e c^2} - 4k_u^2 \eta^2 \right] \tilde{a} - i\,4k_u \eta \tilde{a}' \right) . \tag{4.42}$$

Finally, $\tilde{a}'$ is eliminated by multiplying Eq. (4.40) with $i\,4k_u \eta$ and adding it to Eq. (4.42), and $\tilde{a}$ is eliminated with the help of Eq. (4.38). Using $j_0 = -n_e e \, c$ one finally arrives at the famous third-order differential equation of the high-gain FEL

$$\tilde{E}_x''' + i\,4k_u\eta\tilde{E}_x'' + \left(k_p^2 - 4k_u^2\eta^2\right)\tilde{E}_x' - i\,\Gamma^3\tilde{E}_x = 0. \tag{4.43}$$

Two new coefficients appear in this equation which depend on the beam properties and the layout of the undulator, and which have both the dimension of an inverse length. The first one is called the *gain parameter* Γ, the second one is often called the *space charge parameter* k_p

$$\Gamma = \left[\frac{\mu_0\widehat{K}^2e^2k_un_e}{4\gamma_r^3m_e}\right]^{1/3}, \quad k_p = \sqrt{\frac{2k_u\mu_0n_ee^2c}{\gamma_rm_e\omega_\ell}} = \sqrt{\frac{2\lambda_\ell}{\lambda_u}}\cdot\frac{\omega_p^*}{c}. \tag{4.44}$$

The space charge parameter is related to the plasma frequency ω_p^* in the relativistic electron bunch

$$\omega_p^* = \sqrt{\frac{n_e^*e^2}{\varepsilon_0m_e}} = \sqrt{\frac{n_ee^2}{\gamma_r\varepsilon_0m_e}}. \tag{4.45}$$

(When computing the plasma frequency in the relativistic bunch one has to take into account that the particle density in the rest frame of the bunch is $n_e^* = n_e/\gamma_r$ due to Lorentz expansion of the bunch length). We will see in Chap. 6 that space charge forces can be neglected if k_p is small in comparison with the gain parameter Γ. If in addition the electron beam is on resonance we get the simplest form of the third-order equation, valid for $\eta = 0$ and $k_p = 0$:

$$\tilde{E}_x''' - i\,\Gamma^3\tilde{E}_x = 0\,. \tag{4.46}$$

In the general case, it is instructive to rewrite the third-order equation in the form

$$\frac{\tilde{E}_x'''}{\Gamma^3} + 2i\,\frac{2k_u}{\Gamma}\,\eta\,\frac{\tilde{E}_x''}{\Gamma^2} + \left(\frac{k_p^2}{\Gamma^2} - \left(\frac{2k_u}{\Gamma}\,\eta\right)^2\right)\frac{\tilde{E}_x'}{\Gamma} - i\,\tilde{E}_x = 0\,.$$

The coefficient of the relative energy deviation $\eta = (W - W_r)/W_r$ is $2k_u/\Gamma$. This motivates the introduction of a new quantity which is called the FEL parameter (or Pierce parameter)

$$\rho_{\mathrm{FEL}} = \frac{\Gamma}{2k_u} = \frac{1}{4\pi\sqrt{3}}\cdot\frac{\lambda_u}{L_{g0}}, \tag{4.47}$$

where L_{g0} is the power gain length defined below in Eq. (4.51). With the dimensionless FEL parameter the third-order equation reads

$$\frac{\tilde{E}_x'''}{\Gamma^3} + 2i\,\frac{\eta}{\rho_{\mathrm{FEL}}}\,\frac{\tilde{E}_x''}{\Gamma^2} + \left(\frac{k_p^2}{\Gamma^2} - \left(\frac{\eta}{\rho_{\mathrm{FEL}}}\right)^2\right)\frac{\tilde{E}_x'}{\Gamma} - i\,\tilde{E}_x = 0\,. \tag{4.48}$$

In the next section we will convince ourselves that the third-order equation in its simplest form (4.46) is easy to solve. The general form (4.48) of this equation shows quite clearly that significant changes must be expected if the fractional energy deviation is non-zero and approaches the FEL parameter. This will be studied in Chap. 5. We will demonstrate that the FEL parameter is closely related to two important properties of a high-gain FEL: its bandwidth and its saturation power.

4.8 Analytic Solution of the Third-Order Equation

The linear third-order differential Eq. (4.46) can be solved analytically using the trial function $\tilde{E}_x(z) = A e^{\alpha z}$. Then one obtains $\alpha^3 = i\,\Gamma^3$ with the three solutions

$$\alpha_1 = (i + \sqrt{3})\Gamma/2\,, \quad \alpha_2 = (i - \sqrt{3})\Gamma/2\,, \quad \alpha_3 = -i\Gamma\,. \tag{4.49}$$

The first solution has a positive real part and leads to an exponential growth of the field $\tilde{E}_x(z)$, while the other two eigenvalues correspond to exponentially damped or oscillatory eigenfunctions. For sufficiently large z the power of the light wave grows as

$$P(z) \propto \exp(2\,\Re\{\alpha_1\}\,z) = \exp(\sqrt{3}\Gamma z) \equiv \exp(z/L_{g0})\,, \tag{4.50}$$

where we have defined the *power gain length* by[4]

$$\boxed{L_{g0} = \frac{1}{\sqrt{3}\,\Gamma} = \frac{1}{\sqrt{3}}\left[\frac{4\,\gamma_r^3\,m_e}{\mu_0\,\widehat{K}^2\,e^2\,k_u\,n_e}\right]^{1/3}}\,. \tag{4.51}$$

The index "0" in formula (4.51) indicates that L_{g0} is the idealized gain length of the one-dimensional theory, assuming a mono-energetic beam and neglecting space charge forces. The power gain length L_g of a realistic FEL will in general be larger because beam energy offset and energy spread, space charge, betatron oscillations and optical diffraction tend to weaken the exponential gain with the unfortunate consequence that the undulator length must be increased if one wants to preserve the FEL gain and aims at laser saturation. This will be investigated in Chap. 6.

In the general case $\eta \neq 0$ and $k_p \neq 0$ the linear third-order differential equation is again solved by assuming a z dependence of the form $\exp(\alpha z)$. The resulting cubic equation for the exponent α has three solutions $\alpha_1, \alpha_2, \alpha_3$ which are computed in Sect. 10.4. The general solution of Eq. (4.48) can be written as a linear combination of the three eigenfunctions $V_j(z) = \exp(\alpha_j z)$:

$$\tilde{E}_x(z) = c_1 V_1(z) + c_2 V_2(z) + c_3 V_3(z)\,, \qquad V_j(z) = \exp(\alpha_j z)\,. \tag{4.52}$$

---

[4] Some authors define the *field gain length* which is a factor of two larger than the power gain length.

For the first and second derivative we obtain

$$\tilde{E}_x'(z) = c_1\alpha_1 V_1(z) + c_2\alpha_2 V_2(z) + c_3\alpha_3 V_3(z)\,,$$
$$\tilde{E}_x''(z) = c_1\alpha_1^2 V_1(z) + c_2\alpha_2^2 V_2(z) + c_3\alpha_3^2 V_3(z)\,.$$

Since $V_j(0) = 1$ the coefficients c_j can be computed by specifying the initial conditions for $\tilde{E}_x(z)$, $\tilde{E}_x'(z)$ and $\tilde{E}_x''(z)$ at the beginning of the undulator at $z = 0$. The initial values can be expressed in matrix form by

$$\begin{pmatrix} \tilde{E}_x(0) \\ \tilde{E}_x'(0) \\ \tilde{E}_x''(0) \end{pmatrix} = \boldsymbol{A} \cdot \begin{pmatrix} c_1 \\ c_2 \\ c_3 \end{pmatrix} \quad \text{with} \quad \boldsymbol{A} = \begin{pmatrix} 1 & 1 & 1 \\ \alpha_1 & \alpha_2 & \alpha_3 \\ \alpha_1^2 & \alpha_2^2 & \alpha_3^2 \end{pmatrix}. \tag{4.53}$$

The coefficient vector is then computed as

$$\begin{pmatrix} c_1 \\ c_2 \\ c_3 \end{pmatrix} = \boldsymbol{A}^{-1} \cdot \begin{pmatrix} \tilde{E}_x(0) \\ \tilde{E}_x'(0) \\ \tilde{E}_x''(0) \end{pmatrix}. \tag{4.54}$$

To be more specific we go back to the simple case $\eta = 0$ and $k_p = 0$. Then the eigenvalues are given by Eq. (4.49) and the matrix $\boldsymbol{A}$ and its inverse assume the forms

$$\boldsymbol{A} = \begin{pmatrix} 1 & 1 & 1 \\ (i+\sqrt{3})\Gamma/2 & (i-\sqrt{3})\Gamma/2 & -i\Gamma \\ (i+\sqrt{3})^2\Gamma^2/4 & (i-\sqrt{3})^2\Gamma^2/4 & -\Gamma^2 \end{pmatrix}, \tag{4.55}$$

$$\boldsymbol{A}^{-1} = \frac{1}{3} \cdot \begin{pmatrix} 1 & (\sqrt{3}-i)/(2\Gamma) & (-i\sqrt{3}+1)/(2\Gamma^2) \\ 1 & (-\sqrt{3}-i)/(2\Gamma) & (i\sqrt{3}+1)/(2\Gamma^2) \\ 1 & i/\Gamma & -1/\Gamma^2 \end{pmatrix}. \tag{4.56}$$

Let now the FEL process be started by an incident plane light wave of wavelength λ_ℓ and amplitude E_{in}

$$E_x(z,t) = E_{\text{in}}\cos(k_\ell z - \omega_\ell t) \quad \text{with} \quad k_\ell = \omega_\ell/c = 2\pi/\lambda_\ell\,,$$

which to a good approximation happens when the FEL is seeded by an external laser. The initial density modulation is assumed to be zero: $\tilde{j}_1(0) = 0$, so $\tilde{E}_x'(0) = 0$ from Eq. (4.21). Also the second derivative of the field vanishes at $z = 0$. Hence the initial condition is

$$\begin{pmatrix} \tilde{E}_x(0) \\ \tilde{E}_x'(0) \\ \tilde{E}_x''(0) \end{pmatrix} = \begin{pmatrix} E_{\text{in}} \\ 0 \\ 0 \end{pmatrix}. \tag{4.57}$$

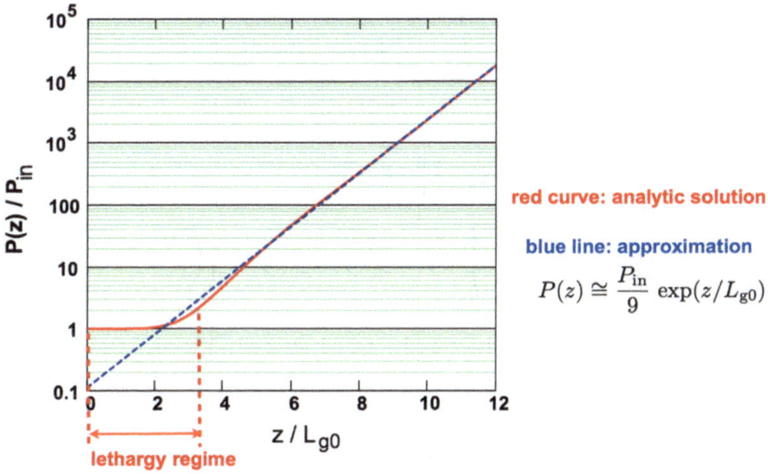

Fig. 4.3 Computed power rise in a high-gain FEL. The lasing process is started by incident *seed radiation*, the input light wave is a plane wave with power P_{in}. The electron beam is mono-energetic and on resonance, i.e. $\eta = 0$. The solid curve shows the normalized FEL power $P(z)/P_{\text{in}}$ as a function of z/L_{g0}. The dashed line is the exponential function $f(z) = (P_{\text{in}}/9)\,\exp(z/L_{\text{g0}})$.

From Eq. (4.54) we find that all three coefficients have the same value, $c_j = E_{\text{in}}/3$. So the field of the FEL wave is

$$\tilde{E}_x(z) = \frac{E_{\text{in}}}{3}\left[\exp((i+\sqrt{3})\Gamma z/2) + \exp((i-\sqrt{3})\Gamma z/2) + \exp(-i\Gamma z)\right]. \quad (4.58)$$

The first term in the square bracket exhibits exponential growth, the second term exponential decay, and the third term oscillates with constant amplitude as a function of the position z along the undulator axis. After a certain distance the first term dominates and the FEL field grows as

$$\left|\tilde{E}_x(z)\right| \approx \frac{E_{\text{in}}}{3}\exp(\sqrt{3}\Gamma z/2) \equiv \frac{E_{\text{in}}}{3}\exp\left(\frac{z}{(2L_{\text{g0}}}\right).$$

The FEL power as function of z is plotted in Fig. 4.3. Due to the presence of the other two terms it stays almost constant in the so-called "lethargy regime" ($0 \le z \lesssim 3\,L_{\text{g0}}$ in the present example) and grows then asymptotically as

$$P(z) \cong \frac{P_{\text{in}}}{9}\exp(z/L_{\text{g0}}) \quad \text{for} \quad z \ge 3\,L_{\text{g0}}\,, \quad (4.59)$$

where P_{in} is the power of the incident seed light wave at $z = 0$. The starting value of the exponential function is one ninth of the incident power P_{in}. This behavior is quite typical for seeded high-gain FELs.

Table 4.1 Comparison of FELs with planar resp. helical undulator

Planar undulator	Helical undulator
$\lambda_\ell = \frac{\lambda_u}{2\gamma^2}\left(1+\frac{K^2}{2}\right), \quad K = \frac{eB_0\lambda_u}{2\pi m_e c}$	$\lambda_\ell = \frac{\lambda_u}{2\gamma^2}(1+K^2), \quad K = \frac{eB_0\lambda_u}{2\pi m_e c}$
$\widehat{K} = K\cdot\left[J_0\left(\frac{K^2}{4+2K^2}\right) - J_1\left(\frac{K^2}{4+2K^2}\right)\right]$	$\widehat{K} \equiv K$
$\frac{d\eta}{dz} = -\frac{e}{m_e c^2\gamma_r}\Re\left\{\left(\frac{\widehat{K}\tilde{E}_x}{2\gamma_r} + \tilde{E}_z\right)e^{i\psi}\right\}$	$\frac{d\eta}{dz} = -\frac{e}{m_e c^2\gamma_r}\Re\left\{\left(\frac{K}{2\gamma_r}(\tilde{E}_x \pm i\tilde{E}_y) + \tilde{E}_z\right)e^{i\psi}\right\}$ $= -\frac{e}{m_e c^2\gamma_r}\Re\left\{\left(\frac{K\tilde{E}_x}{\gamma_r} + \tilde{E}_z\right)e^{i\psi}\right\}$
$\frac{d\tilde{E}_x}{dz} = -\frac{\mu_0 c\widehat{K}}{4\gamma}\cdot\tilde{j}_1$	$\frac{d\tilde{E}_x}{dz} = -\frac{\mu_0 cK}{4\gamma}\cdot\tilde{j}_1$
$\tilde{E}_z(z) = i\,\frac{4\gamma c}{\omega_\ell\widehat{K}}\cdot\frac{d\tilde{E}_x}{dz}$	$\tilde{E}_z(z) = i\,\frac{4\gamma c}{\omega_\ell K}\cdot\frac{d\tilde{E}_x}{dz}$
$\Gamma = \left[\frac{\mu_0\widehat{K}^2 e^2 k_u n_e}{4\gamma_r^3 m_e}\right]^{1/3}$	$\Gamma = \left[\frac{\mu_0 K^2 e^2 k_u n_e}{2\gamma_r^3 m_e}\right]^{1/3}$

4.9 High-Gain FEL with Helical Undulator

We present here without proof a short description of the electron motion in a helical undulator and the emission of circularly polarized radiation. Some of the results are derived in Sect. 10.1. A helical undulator can be thought of as a long dipole magnet that is twisted about its axis. The idealized field (valid near the axis) for a right-handed screw sense is

$$\boldsymbol{B} = -B_0\left[\cos(k_u z)\boldsymbol{e}_x + \sin(k_u z)\boldsymbol{e}_y\right], \tag{4.60}$$

where $\boldsymbol{e}_x$ and $\boldsymbol{e}_y$ are the unit vectors in x and y direction, respectively. Note that this field is only approximate and does not obey the Maxwell equations. The electron moves on a helical trajectory of radius

$$r_{\text{hel}} = \frac{K}{\gamma k_u}$$

with a constant longitudinal speed, given by

$$v_z \approx v_0 \equiv c\left[1 - \frac{1}{2\gamma^2}\left(1+K^2\right)\right]. \tag{4.61}$$

The radiation produced in the undulator field (4.60) has positive helicity and its electric field vector can be written in the form (see Sect. 10.1)

$$\boldsymbol{E}(z,t) = E_0 \left[\cos(k_\ell z - \omega_\ell t)\boldsymbol{e}_x - \sin(k_\ell z - \omega_\ell t)\boldsymbol{e}_y\right]. \tag{4.62}$$

The wavelength is

$$\lambda_\ell = \frac{\lambda_u}{2\gamma^2}\left(1 + K^2\right). \tag{4.63}$$

If the helical magnet is twisted like a left-handed screw the circularly polarized radiation will have negative helicity.

In Table 4.1 we compare important high-gain FEL equations which are different for planar and helical undulators. The main changes concern the wavelength and the undulator parameter: $\widehat{K}$ has to be replaced by K when going from a planar to a helical undulator. Moreover one has to keep in mind that the electric vector of the radiation field has two components which couple both to the electron and double the energy transfer. They are related by $\tilde{E}_y = \mp i \tilde{E}_x$. The space charge parameter k_p remains invariant, and the third-order equation retains its form (4.48) if the correct value of the gain parameter Γ is used.

References

1. FEL at Jefferson Lab, see www.jlab.org/FEL/
2. Y.V. Shvyd'ko et al., High-reflectivity high-resolution X-ray crystal optics with diamond. Nat. Phys. **6**, 196 (2010)
3. V. Ayvazyan et al., Generation of GW radiation pulses from a VUV free-electron laser operating in the femtosecond regime. Phys. Rev. Lett. **88**, 104802 (2002)
4. D.A. Edwards, M.J. Syphers, *An Introduction to the Physics of High Energy Accelerators* (John Wiley, New York, 1993)
5. Z. Huang, K.-J. Kim, Review of x-ray free-electron laser theory. Phys. Rev. ST Accel. Beams **10**, 034801 (2007)

Chapter 5
Applications of the High-Gain FEL Equations

In the first part of this chapter we want to exploit the third-order differential equation for the amplitude of the FEL wave in order to obtain a deeper understanding of the properties and peculiarities of high-gain free-electron lasers. It will turn out that a remarkable number of predictions can be deduced from the analytic description of the high-gain FEL although we have made rather simplifying assumptions in deriving Eq. (4.48), for example by neglecting any dependencies on the transverse coordinates x and y. The coupled first-order equations are evaluated in the second part. They contain even more physics. We will apply them to study the saturation regime and the evolution of microbunching. In the model calculations discussed in this chapter we investigate the amplifier mode of the high-gain FEL and consider either a perfectly uniform initial particle distribution, in which case the lasing process has to be started by an incident light wave, or we assume a density modulation of the electron bunch which is periodic in the optical wavelength. Then no "seed light" is needed. The startup from "noise" via the SASE mechanism will be investigated in Chap. 7.

5.1 Gain Function of the High-Gain FEL

5.1.1 Third-Order Equation with Energy Detuning

The third-order equation (4.48) has been solved in the previous chapter for the simplest case of a mono-energetic electron beam ($\sigma_\eta = \sigma_W / W_r = 0$) whose energy conforms with the FEL resonance condition $W = W_r = \gamma_r m_e c^2$. The third-order equation can also be solved for a beam energy $W \neq W_r$. The procedure is as follows. In the first step the three complex eigenvalues α_j are calculated using the methods described in Sect. 10.4. These eigenvalues are functions of the energy detuning $\eta = (W - W_r)/W_r$, see Eq. (10.93). Once they are known, the general solution of Eq. (4.48) can be written in the form

P. Schmüser et al., *Free-Electron Lasers in the Ultraviolet and X-Ray Regime*,
Springer Tracts in Modern Physics 258, DOI: 10.1007/978-3-319-04081-3_5,

$$\tilde{E}_x(\eta, z) = \sum_{j=1}^{3} c_j(\eta) \exp(\alpha_j(\eta) z) \,. \tag{5.1}$$

The field $\tilde{E}_x$ inside the undulator magnet depends implicitly on the relative energy deviation η through the eigenvalues $\alpha_j(\eta)$. The coefficients c_j are determined by the initial conditions, using Eqs. (4.53) and (4.54):

$$\begin{pmatrix} c_1 \\ c_2 \\ c_3 \end{pmatrix} = \begin{pmatrix} 1 & 1 & 1 \\ \alpha_1 & \alpha_2 & \alpha_3 \\ \alpha_1^2 & \alpha_2^2 & \alpha_3^2 \end{pmatrix}^{-1} \cdot \begin{pmatrix} \tilde{E}_x(0) \\ \tilde{E}_x'(0) \\ \tilde{E}_x''(0) \end{pmatrix} . \tag{5.2}$$

From this equation it is clear that the coefficients depend on η as well.

Let again the lasing process be started by an incident plane wave of amplitude $E_{\rm in}$. The gain function $G(\eta, z)$ of an FEL amplifier depends on the relative energy deviation η and the position z in the undulator

$$G(\eta, z) = \left| \frac{\tilde{E}_x(\eta, z)}{E_{\rm in}} \right|^2 - 1 \,. \tag{5.3}$$

(Compare the discussion of the FEL gain function in Chap. 3).

5.1.2 Short Undulator: Low-Gain Limit

We have seen in Fig. 4.3 of the previous chapter that the FEL power stays almost constant in the lethargy regime $0 \le z \lesssim 3\,L_{g0}$. Therefore we expect the low-gain FEL theory to be applicable here. This is indeed the case. As an illustration we consider a short undulator magnet with a length $L_u = L_{g0}$. The gain curve computed according to Eq. (5.3) is compared in Fig. 5.1 with the gain curve obtained in the low-gain theory, using the Madey theorem Eq. (3.26). An almost perfect agreement is observed, both in the shape of the gain curve and also in its magnitude. This demonstrates that the low-gain FEL theory is the limiting case of the more general high-gain theory for short undulator magnets. The mathematical proof of this statement is presented in Sect. 10.2.2.

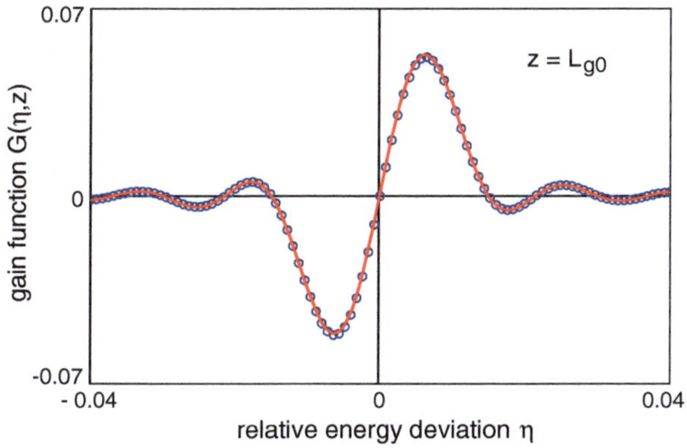

Fig. 5.1 Comparison of the high-gain and the low-gain FEL theory for a short undulator magnet with a length $L_u = L_{g0}$. The continuous curve shows the gain function $G(\eta, z = L_{g0})$ according to Eq. (5.3) of the high-gain FEL theory. The circles show the gain function according to the low-gain FEL theory, using the Madey theorem Eq. (3.26) with the replacement $N_u\lambda_u \to L_{g0}$. Note that the maximum value of the gain function is only 0.05, so one is indeed in the low-gain regime.

5.1.3 Long Undulator: High-Gain Regime

For undulator magnets that are longer than several gain lengths we obtain significant differences between the low-gain and the high-gain theory. This is demonstrated in Fig. 5.2 where the gain function is shown at various longitudinal positions inside a long undulator ($L_u \gg L_{g0}$). In the next section we will see that the width of the gain curve is closely related to the FEL parameter, which is the reason why $G(\eta, z)$ is plotted as a function of $\eta/\rho_{\rm FEL}$. For a long undulator, the high-gain FEL features a much larger amplification than the low-gain FEL. A very interesting observation is that the maximum amplification happens close to the point $\eta = 0$, i.e. on resonance, where the gain function vanishes in the low-gain theory.

5.2 FEL Bandwidth

The width of the gain curves $G(\eta, z)$ shown in Fig. 5.2 shrinks with increasing undulator length. Which quantity determines the bandwidth of the high-gain FEL amplifier? A good measure for the bandwidth turns out to be the dimensionless FEL parameter defined in the previous chapter

$$\rho_{\rm FEL} = \frac{1}{4\pi\sqrt{3}} \cdot \frac{\lambda_u}{L_{g0}} .$$

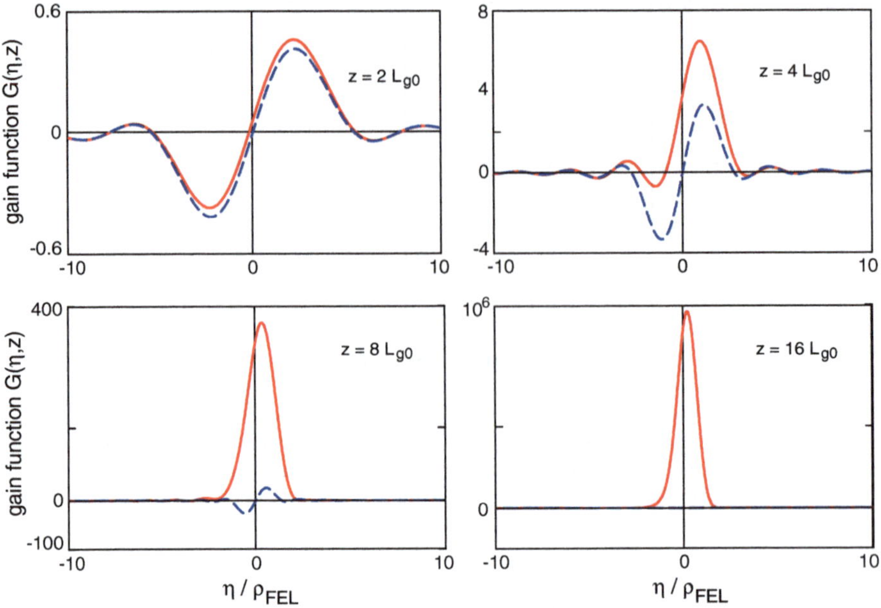

Fig. 5.2 FEL gain function $G(\eta, z)$ plotted versus η/ρ_{FEL} at various positions in a long undulator: $z = 2\,L_{g0}, 4\,L_{g0}, 8\,L_{g0}, 16\,L_{g0}$. *Continuous red curves*: computation according to the high-gain theory. *Dashed blue curves*: gain function according to the low-gain theory, using the Madey theorem (3.26). Note the very different vertical scales in the figures.

The numerical value of the FEL parameter is typically in the 10^{-3} range. One observes that the FEL gain drops significantly when the relative energy deviation η exceeds the FEL parameter. The full width at half maximum (FWHM) of the gain curves of Fig. 5.2 is about $2\,\rho_{\mathrm{FEL}}$ at $z = 4\,L_{g0}$ and drops to about $1\,\rho_{\mathrm{FEL}}$ at $z = 16\,L_{g0}$. These observations show that the high-gain FEL acts as a narrow-band amplifier of variable bandwidth.

5.2.1 Detuning Parameter

The third-order equation describes the FEL process for a given combination of light frequency ω_ℓ and beam energy $W = \gamma m_e c^2$. The eigenvalues α_j depend on the *detuning*. Detuning means a deviation from the resonance condition. We distinguish two important cases:

(1) FEL seeding with a single frequency

In a monochromatically seeded FEL, which we have extensively studied in the previous sections, the incident seed light of frequency ω_ℓ defines the resonant frequency

$$\omega_r \equiv \omega_\ell \,, \tag{5.4}$$

but the electron energy $W = \gamma m_e c^2$ may be different from the resonant energy $W_r = \gamma_r m_e c^2$. In this case the detuning parameter is a function of γ

$$\eta = \eta(\gamma) = \frac{\gamma - \gamma_r}{\gamma_r} \quad \text{with} \quad \gamma_r = \sqrt{\frac{\omega_\ell}{2c\,k_u}\left(1 + \frac{K^2}{2}\right)} \,. \tag{5.5}$$

(2) Seeding with different frequencies or SASE FEL

Suppose several stimulation frequencies $\omega_1, \omega_2, \ldots$ are simultaneously present, either in an incident seed wave or in the Fourier spectrum of the bunch charge distribution. The latter case applies for the SASE FEL. Now it is appropriate to take the electron beam energy W as our reference (we disregard energy spread at this point)

$$W \equiv W_r = \gamma_r m_e c^2 \,. \tag{5.6}$$

The corresponding reference frequency is

$$\omega_r = \frac{2\,\gamma_r^2 c\,k_u}{1 + K^2/2} \,, \tag{5.7}$$

and we can define the detuning parameter as a function of frequency

$$\eta = \eta(\omega) = -\frac{\omega - \omega_r}{2\,\omega_r} \,. \tag{5.8}$$

The factor of two in the denominator is due to the proportionality $\omega_r \propto \gamma_r^2$, and the minus sign comes in because an electron energy higher than the resonance energy in a seeded FEL is equivalent to a light frequency lower than the resonance frequency in a SASE FEL. With Eq. (5.8) we have basically translated a frequency deviation of the FEL wave into an equivalent energy deviation of the electron beam.

The coefficients $c_j(\eta_1), c_j(\eta_2), \ldots$ of the eigenfunction expansion are now calculated individually by applying Eq. (5.2) for each detuning parameter $\eta_\nu = \eta(\omega_\nu)$. The resulting field is the superposition of the individual contributions

$$\begin{aligned} E_x(z,t) &= \Re\left\{\sum_\nu \tilde{E}_x^{(\nu)}(z)\exp(i\omega_\nu[z/c - t])\right\} \\ &= \Re\left\{\exp(i\omega_r[z/c - t])\sum_\nu \tilde{E}_x^{(\nu)}(z)\exp(i\Delta\omega_\nu[z/c - t])\right\}, \end{aligned} \tag{5.9}$$

where $\Delta\omega_\nu = \omega_\nu - \omega_r$ is the respective frequency deviation from resonance. The time variable can be eliminated using the relation (compare Sect. 10.2.4)

$$\zeta(z,t) = z - \bar{\beta} c t$$

which means that the field can be considered as being a function of the position z in the undulator and the internal coordinate ζ in the bunch:

$$E_x(z,t) = \Re\left\{\exp(i\omega_r[z/c - t)]) \sum_{\nu} \widehat{E}^{\nu}(z, \zeta(z,t))\right\} \tag{5.10}$$

with the complex field amplitudes

$$\widehat{E}^{\nu}(z,\zeta) = \tilde{E}_x^{\nu}(z)\exp(i\Delta\omega_{\nu}[z/c + (\zeta - z)/(\bar{\beta} c)]) \,. \tag{5.11}$$

The current density can be computed by means of Eq. (4.21). The method sketched here will be applied in Sect. 5.5 to investigate under which circumstances the superposition principle is applicable in a high-gain FEL.

5.2.2 *Analytical Determination of the FEL Bandwidth*

In order to investigate the relevance of the FEL parameter in a quantitative way we determine the eigenvalues α_j for the case that the beam energy W differs from the resonance energy W_r. In the special case of a mono-energetic beam ($\sigma_\eta = 0$), and for negligible space charge ($k_p = 0$), the three complex eigenvalues $\alpha_j(\eta)$ are computed analytically in Sect. 10.4, see Eqs. (10.93), (10.94). In the exponential regime the eigenfunction $\exp(\alpha_1 z)$ dominates and the real part of α_1 determines the growth rate of the field $\tilde{E}_x$. The functional dependence $\Re[\alpha_1(\eta)]$ is discussed in Sect. 10.4. In the vicinity of $\eta = 0$ the real part of $\alpha_1(\eta)$ can be expanded in a Taylor series, see Eq. (10.95):

$$\Re[\alpha_1(\eta)] \approx \frac{1}{2L_{g0}}\left(1 - \frac{\eta^2}{9\,\rho_{\rm FEL}^2}\right). \tag{5.12}$$

From this follows that the gain function in the exponential-growth regime can be approximated by the product of an exponential function and a Gaussian[1]

$$G(\eta, z) \propto \exp(z/L_{g0})\,\exp\left(-\frac{\eta^2 z}{9\,\rho_{\rm FEL}^2\,L_{g0}}\right).$$

[1] In reality the Gaussian is not centered at $\eta = 0$ but shifted to slightly positive values. This shift is due to the energy dependence of the coefficients $c_j(\eta)$ and can be ignored here because it has a negligible effect on the bandwidth. The term "-1" in Eq. (5.3) can be dropped in the high-gain regime where $|\tilde{E}_x(\eta, z)|^2/E_{\rm in}^2 \gg 1$.

Using Eq. (5.8), the gain function is rewritten as a function of ω

$$G(\omega, z) \propto \exp(z/L_{g0}) \exp\left(-\frac{(\omega - \omega_r)^2}{2\sigma_\omega^2}\right). \tag{5.13}$$

The rms frequency bandwidth depends on the distance z traveled in the undulator

$$\sigma_\omega = \sigma_\omega(z) = 3\sqrt{2}\,\rho_{\text{FEL}}\,\omega_\ell\sqrt{\frac{L_{g0}}{z}}\,. \tag{5.14}$$

Formula (5.13) is approximately valid between $z \gtrsim 8\,L_{g0}$ and the beginning of saturation. The bandwidth (5.14) will be used in Chap. 7 to compute the power of a SASE FEL.

5.3 FEL Startup by a Periodically Modulated Electron Beam Current

Now we want to demonstrate that the FEL process can also be initiated by a periodic charge density modulation in the electron beam, instead of an incident light wave for seeding. Hence we consider a beam current which is modulated periodically in the ponderomotive phase variable ψ according to Eq. (4.4)

$$j_z = j_0 + \tilde{j}_1(z)\exp(i\psi)\,.$$

In the previous chapter we have shown that this leads to a non-vanishing derivative of the transverse electric field, see Eq. (4.21). At the entrance of the undulator at $z = 0$ one gets

$$E_0 \equiv \tilde{E}_x(0) = 0 \quad \text{and} \quad E_0^{'} \equiv \frac{d\tilde{E}_x}{dz}(0) = -\frac{\mu_0 c\widehat{K}}{4\gamma_r}\cdot\tilde{j}_1(0)\,.$$

The second derivative is

$$E_0^{''} = -\frac{\mu_0 c\widehat{K}}{4\gamma_r}\cdot\tilde{j}_1^{'}(0)\,,$$

so the question arises how to compute the derivative of the current density. For that purpose we use two of the coupled equations (4.31):

$$\tilde{j}_1(z) = j_0\frac{2}{N}\sum_{n=1}^{N}\exp(-i\psi_n(z)) \quad \text{and} \quad \frac{d\psi_n}{dz} = 2k_u\eta_n\,.$$

Under the assumption that initially all particles have the same energy

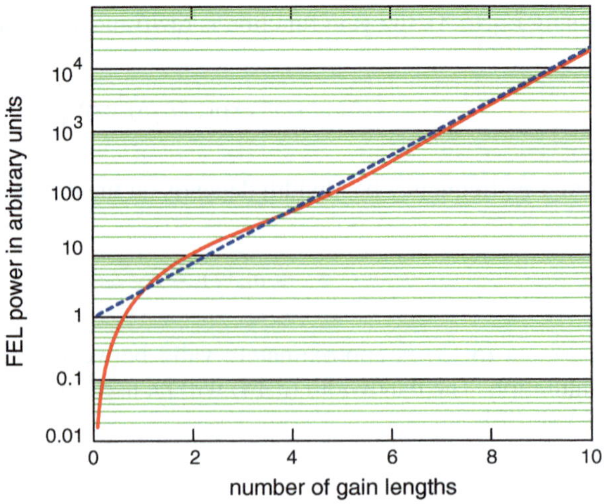

Fig. 5.3 Power rise in a high-gain FEL which is started by a periodic initial electron charge density modulation with the period λ_ℓ. The electron beam is assumed to be mono-energetic and on resonance, i.e. $W = W_r$ and $\eta = 0$. Shown as a *dotted line* is the exponential function $f(z) = \exp(z/L_{g0})$.

$$\eta_1(0) = \eta_2(0) = \ldots \eta_N(0) \equiv \eta$$

we obtain

$$\tilde{j}_1'(0) = \left[\sum_{n=1}^{N} \frac{d\tilde{j}_1}{d\psi_n}\frac{d\psi_n}{dz}\right]_{z=0} = -i\ \underbrace{j_0\frac{2}{N}\sum_{n=1}^{N}\exp(-i\psi_n(0))}_{\tilde{j}_1(0)}\ 2\,k_u\,\eta$$

$$= -i\,2\,k_u\,\eta\,\tilde{j}_1(0)\,.$$

Therefore

$$E_0'' = i\,2k_u\eta\,\frac{\mu_0 c\widehat{K}}{4\gamma_r}\cdot\tilde{j}_1(0)\,,$$

and the initial condition of the third-order equation reads

$$\begin{pmatrix}\tilde{E}_x(0)\\ \tilde{E}_x'(0)\\ \tilde{E}_x''(0)\end{pmatrix} = \begin{pmatrix}0\\ E_0'\\ E_0''\end{pmatrix} = \begin{pmatrix}0\\ -1\\ i\,2k_u\eta\end{pmatrix}\frac{\mu_0 c\,\widehat{K}}{4\gamma_r}\,\tilde{j}_1(0)\,. \tag{5.15}$$

It is straightforward to compute the coefficients c_j of the eigenfunction expansion

$$\tilde{E}_z(z) = \sum_j c_j\exp(\alpha_j z)$$

with the help of Eq. (5.2). The resulting FEL power as a function of undulator length is shown in Fig. 5.3. At the entrance of the undulator ($z = 0$) the radiation power is of course zero but it rises rapidly with increasing undulator length and approaches the exponential function $\exp(z/L_{g0})$ after about two gain lengths.

A third type of stimulation of the high-gain FEL process is a periodic energy modulation of the incident electron beam.

5.4 Laser Saturation

The exponential growth cannot continue indefinitely. The beam energy decreases due to the energy loss by radiation, and the modulated current density $\tilde{j}_1$ becomes eventually comparable in magnitude to the dc current density j_0. Moreover, the particles begin to move into the phase space region where energy is taken out of the light wave.

The numerical solution of the coupled differential equations (4.31) will now be applied to study FEL saturation. In this calculation we use typical parameters of the extreme-ultraviolet free-electron laser FLASH (see Chap. 8): Lorentz factor $\gamma = 1000$, number of electrons $N_e = 10^9$, bunch length 100 fs, rms bunch radius 0.07 mm. This leads to the following values of the relevant FEL quantities: peak current 1600 A, 1D gain length $L_{g0} = 0.5$ m, space charge parameter $k_p = 0.24\,\mathrm{m}^{-1}$, FEL parameter $\rho_{\mathrm{FEL}} = 0.003$. To reach saturation at about 20 gain lengths the field of the seed light wave is chosen as $E_{\mathrm{in}} = 5$ MV/m. An important practical consideration is that the seed laser beam has to be well focused over a large part of the lethargy region, typically over at least one gain length, to initiate the FEL gain process. When the exponential amplification sets in, the seed laser beam is no longer needed.

The computed power rise is shown in Fig. 5.4. For $0 \leq z \leq 16\,L_{g0}$ one obtains perfect agreement with the eigenfunction approach of the third-order equation. In the saturation regime the figure displays an oscillatory behavior of the FEL power which means that energy is pumped back and forth between the electron beam and the light wave. This will be studied in more detail in Chap. 6. An interesting observation is that different input powers of the seeding wave lead to the same saturation level. This is demonstrated in Fig. 5.5.

The radiation power at saturation can be very roughly estimated as follows. We assume full modulation, i.e. $\left|\tilde{j}_1\right| \approx |j_0|$. The major part of the intensity is generated in the last section of the exponential regime. The field amplitude at saturation is approximately given by the slope of the field gain curve, multiplied with the field gain length, which is twice as large as the power gain length

$$\left|\tilde{E}_x\right|_{\mathrm{sat}} \approx \left|\frac{d\tilde{E}_x}{dz}\right| \cdot 2\,L_{g0} = \frac{\mu_0 c\,\widehat{K}}{4\gamma_r}\,|j_0|\,2\,L_{g0}\,.$$

The saturation power is

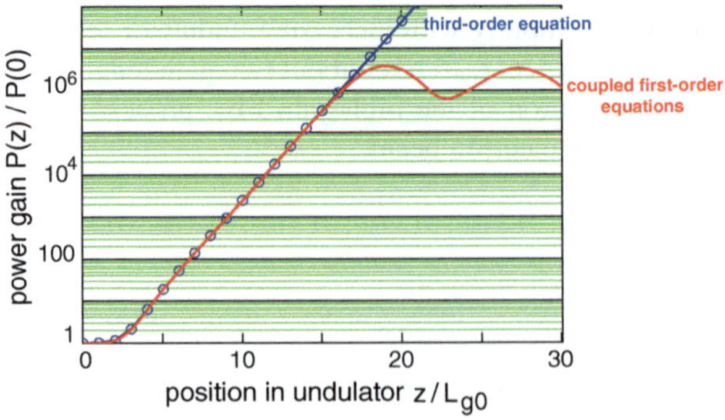

Fig. 5.4 Computed power $P(z)/P_{\rm in}$ as a function of z/L_{g0} in a seeded FEL with the parameters given in the text. The *red curve* shows the result of the numerical integration of the coupled first-order equations. The *circles* have been computed from the third-order equation (4.48). Here the exponential rise continues beyond 16 gain lengths, however it must be kept in mind that Eq. (4.48) is no longer valid in the saturation regime where the modulated current density $\tilde{j}_1$ becomes comparable in magnitude to the dc current density j_0.

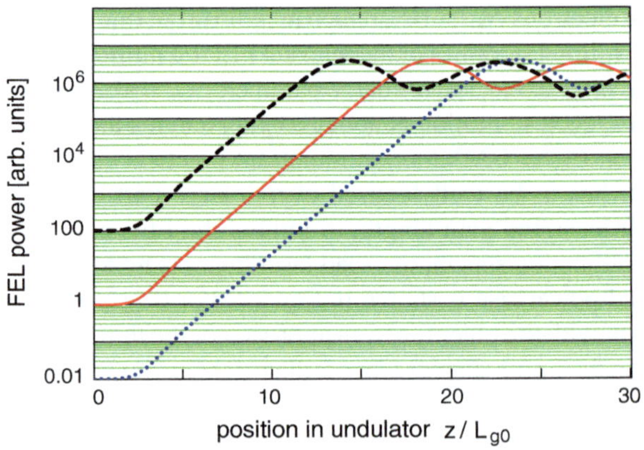

Fig. 5.5 Computed power rise in a high-gain FEL, resulting from the numerical integration of the equations (4.31), for three different power levels of the incident seeding wave. *Red curve*: $P_{\rm in} \equiv P_0 = 1\,$kW, *dashed black curve*: $P_{\rm in} = 100\,P_0$, *dotted blue curve*: $P_{\rm in} = 0.01\,P_0$. In the linear regime, comprising the lethargy and the exponential gain regime, the curves differ by factors of 100 which means that the FEL output power depends here linearly on the input power. The saturation level, however, is the same in all three cases, so the saturated FEL is definitely in the nonlinear regime.

$$P_{\rm sat} = \frac{c\varepsilon_0}{2}\left|\tilde{E}_x\right|^2_{\rm sat} A_b \approx \frac{c\varepsilon_0}{2}\left(\frac{\mu_0 c\,\widehat{K}}{4\gamma_r}\right)^2 \frac{I_0^2}{A_b}\,4\,L_{g0}^2\,,$$

where A_b is the beam cross section and $I_0 = |j_0|\,A_b$ is the magnitude of the dc beam current. Introducing the FEL parameter and the power contained in the electron beam

$$P_{\text{beam}} = \frac{\gamma_r m_e c^2 I_0}{e} \tag{5.16}$$

one finds a rule of thumb for estimating the FEL power at saturation

$$P_{\text{sat}} \approx \rho_{\text{FEL}} \, P_{\text{beam}} \, . \tag{5.17}$$

The FEL power at saturation is thus in the order of 0.1 % of the electron beam power. For typical beam parameters of FLASH one obtains a saturation power of several GW. A better estimate of the saturation power is presented in the next chapter, see Eq. (6.27).

5.5 Linear and Nonlinear Regime of a High-Gain FEL

The free-electron laser is an intrinsically nonlinear system. A linearization has been achieved in Chap. 4 by eliminating the single-particle coordinates. We have derived the linear third-order differential equation (4.48) for the amplitude of the light wave. The solutions of a linear differential equation depend linearly on the initial conditions. The same applies for the integro-differential equation (10.51), derived in Sect. 10.2.1, which is more general than Eq. (4.48) as it can also be used for a beam with energy spread. The linear regime of a high-gain FEL amplifier is characterized by the following features:

(1) The response (output electric field) is proportional to the stimulation (the input field).
(2) The superposition principle applies: the response to a superposition of different stimulations is equal to the superposition of the responses to the individual stimulations.

An illustration of the linear relationship between the input field E_{in} of a monochromatic seed wave and the FEL output field $\tilde{E}_x$ has already been presented in Fig. 5.5 where this linearity is observed over a wide range of z values. The linear relationship is completely lost in the saturation region.

Now we give an example for the superposition principle and study the response of the FEL to a seed radiation containing two different frequencies.

According to Sect. 5.2.1 the electron beam energy W defines now the reference frequency:

$$W \equiv W_r = \gamma_r m_e c^2, \;\; \omega_r = \frac{2\gamma_r^2 c \, k_u}{1 + K^2/2} \, .$$

We choose symmetrically detuned frequencies for our seed waves

$$\omega_a = \omega_r(1 - \rho_{\text{FEL}}) \, , \omega_b = \omega_r(1 + \rho_{\text{FEL}})$$

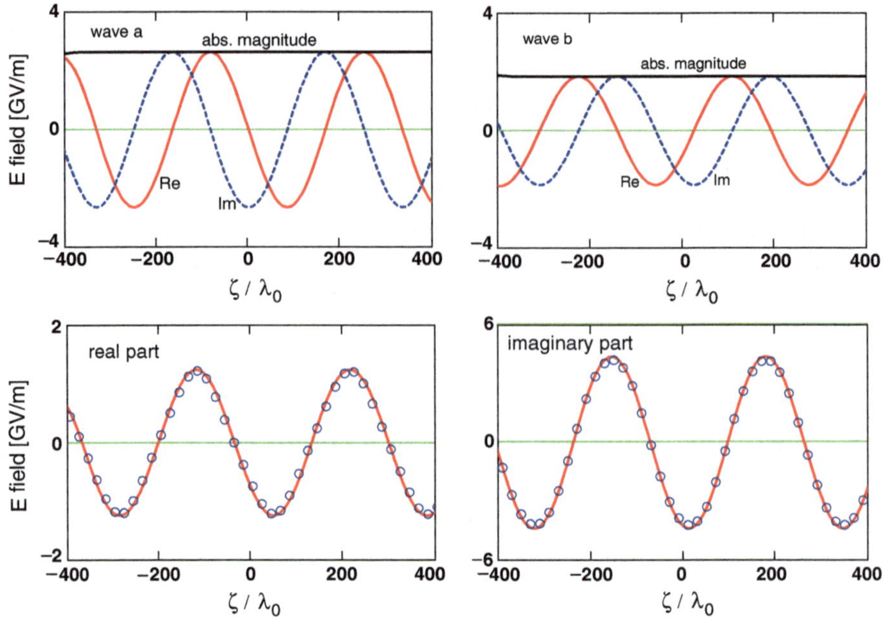

Fig. 5.6 Verification of the superposition principle in the linear regime at $z = 15\,L_{g0}$. *Top left graph*: real part, imaginary part and absolute magnitude of the field $\widehat{E}^a(z,\zeta)$ of frequency ω_a, plotted versus ζ/λ_0 (λ_0 denotes the FEL wavelength corresponding to the resonance energy). *Top right graph*: real part, imaginary part and absolute magnitude of the field $\widehat{E}^b(z,\zeta)$ of frequency ω_b. *Bottom left graph*: the sum of the real parts $\Re\{\widehat{E}^a(z,\zeta)+\widehat{E}^b(z,\zeta)\}$ of the individual responses is shown as a *solid red curve*, the real part $\Re\{\widehat{E}^s(z,\zeta)\}$ of the sum-response (5.19) is shown by the *blue circles*. *Bottom right graph*: same for the imaginary parts.

with equal initial amplitudes, $E^a_{\text{in}} = E^b_{\text{in}} = 5\,\text{MV/m}$. The coupled first-order equations (10.63a, 10.63b, 10.63c, 10.63d) in the non-periodic model are used to compute the individual response functions. We introduce the complex field amplitudes according to Eq. (5.11)

$$\widehat{E}^a(z,\zeta) = \tilde{E}^a_x(z)\exp(i\,\Delta\omega_a[z/c + (\zeta - z)/(\bar{\beta}\,c)])\,, \qquad (5.18)$$
$$\widehat{E}^b(z,\zeta) = \tilde{E}^b_x(z)\exp(i\,\Delta\omega_b[z/c + (\zeta - z)/(\bar{\beta}\,c)])\,,$$

which are functions of the position z along the undulator and the position ζ inside the bunch. In the same manner one computes the sum-response function $\widehat{E}^s(z,\zeta)$ corresponding to the simultaneous excitation with two frequencies and the initial field $E^s_{\text{in}} = E^a_{\text{in}} + E^b_{\text{in}}$. A bunch with a flat charge profile and a length of 1000 optical periods has been used in the computation, but for the sake of clarity the effects at the bunch head and tail are ignored here. The result obtained in the linear regime at $z = 15\,L_{g0}$ is depicted in Fig. 5.6. One finds that the equation

$$\widehat{E}^s(z,\zeta) = \widehat{E}^a(z,\zeta) + \widehat{E}^b(z,\zeta) \qquad (5.19)$$

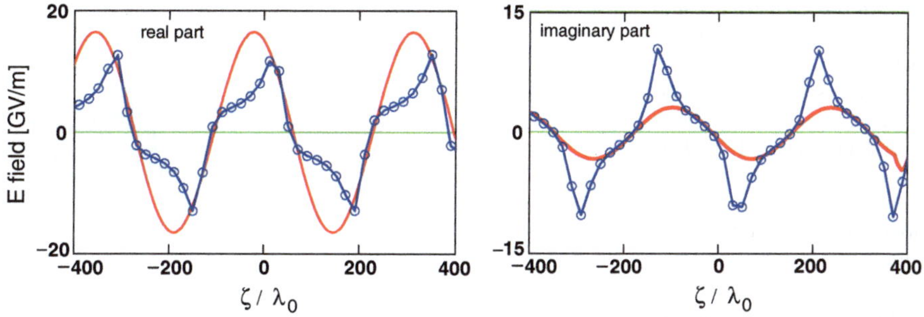

Fig. 5.7 Violation of the superposition principle in the nonlinear regime at $z = 20\,L_{g0}$. *Left graph:* the sum of the real parts $\Re\{\widehat{E}^a(z,\zeta) + \widehat{E}^b(z,\zeta)\}$ of the individual responses is shown as a *solid red curve*, the real part of the sum-response $\Re\{\widehat{E}^s(z,\zeta)\}$ is shown by the *blue circles*. *Right graph:* comparison of the imaginary parts.

is satisfied both for the real part and the imaginary part. Incidentally, the curves shown in Fig. 5.6 can also be computed using Eq. (5.9), but only in the flat-top region of the bunch and not at the head or the tail.

A completely different behavior is found in the nonlinear regime at $z = 20\,L_{g0}$, see Fig. 5.7. Although the individual excitations lead still to almost sinusoidal responses in this example, the sum-response is strongly distorted and the superposition principle is badly violated. Equation (5.9) is obviously not applicable in the nonlinear regime.

The Figs. 5.6 and 5.7 reveal that the amplitude of the field E^a is larger than that of E^b in spite of the symmetrical detuning. This asymmetry is due to an asymmetric shape of the curve $\Re\{\alpha_1(\eta)\}$, see Fig. 6.1 in the next chapter.

In summary one can state that the third-order differential equation is well suited for the description of the linear regime of the high-gain FEL, comprising the lethargy and the exponential-growth regime. This equation breaks down in the nonlinear regime where the laser power goes into saturation. The coupled first-order equations encompass both the linear and the nonlinear regime.

5.6 Simulation of Microbunching

5.6.1 Evolution of Phase Space Structures

In our next example we demonstrate that the periodic form (4.31) of the coupled first-order equations can be used to study the microbunching effect. Again we use typical parameters of FLASH. As initial condition for the numerical simulation we take a beam whose mean electron energy W_0 is equal to the resonance energy W_r, hence $\eta_0 = 0$. A total of $N = 50000$ test particles are distributed uniformly in the initial ponderomotive phase variable over the interval $-3\pi/2 \leq \psi \leq +\pi/2$. The

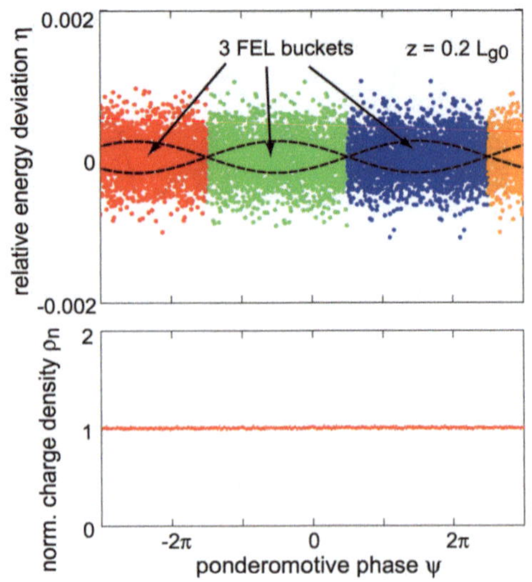

Fig. 5.8 *Top*: Distribution of particles in the (ψ, η) phase space at $z = 0.2\, L_{g0}$. The separatrix is indicated by the dashed curves. Shown are several adjacent FEL buckets. *Bottom*: Normalized charge density $\rho_n(\psi) = |\tilde{\rho}(\psi)| / \rho_0$, plotted as a function of ψ. Remember that the ponderomotive phase ψ is related to the internal bunch coordinate: $\zeta = (\psi + \pi/2)\, \lambda_\ell/(2\pi)$.

distribution is periodically continued to the full range $-\infty < \psi < \infty$, as assumed in the periodic model. A randomly distributed Gaussian energy spread is imposed with $\sigma_\eta = 0.1\, \rho_{\rm FEL}$. The FEL process is started by a seed laser beam with $E_{\rm in} = 5\,$MV/m.

In Fig. 5.8 we show the distribution of the particles in the phase space (ψ, η) shortly after the beginning of the undulator, together with the projection onto the ψ axis. The normalized charge density $\rho_n(\psi, z) = |\tilde{\rho}(\psi, z)| / \rho_0$ shows no significant deviation from unity. Also the energy modulation of the electron bunch is very small and well below the rms energy spread σ_η. The FEL buckets are centered at nearly the same phase values as they are at the beginning of the undulator at $z = 0$, compare Eq. (3.24)

$$\psi_b(0.2\, L_{g0}) \approx \psi_b(0) = -\pi/2 \pm n\, 2\pi\,.$$

The equation of the separatrix can be adopted from the low-gain case, compare Eq. (3.23):

$$\eta_{\rm sep}(\psi) = \pm \sqrt{\frac{e\, \left|\tilde{E}_x(z)\right|\, \widehat{K}}{k_u m_e c^2 \gamma_r^2}} \cos\left(\frac{\psi - \psi_b(z)}{2}\right). \tag{5.20}$$

Due to the imposed beam energy spread of $\sigma_\eta = 0.1\, \rho_{\rm FEL}$, many particles are initially outside the FEL bucket, which is the area enclosed by the separatrix. With increasing distance z traveled in the undulator, the height of the separatrix grows with the square root of the FEL field, according to Eq. (5.20), and in the exponential growth regime most particles are eventually captured in the FEL buckets. This can be seen in the sequence of graphs in Fig. 5.9. Another important observation can be made. The FEL buckets are not stationary but move with increasing z towards smaller phase values. The bucket-center phase $\psi_b(z)$ will be determined in the next section.

Strong changes in the phase space pictures and the charge density distributions become visible after about 10 gain lengths. At $z \geq 12\, L_{g0}$ one observes the evolution

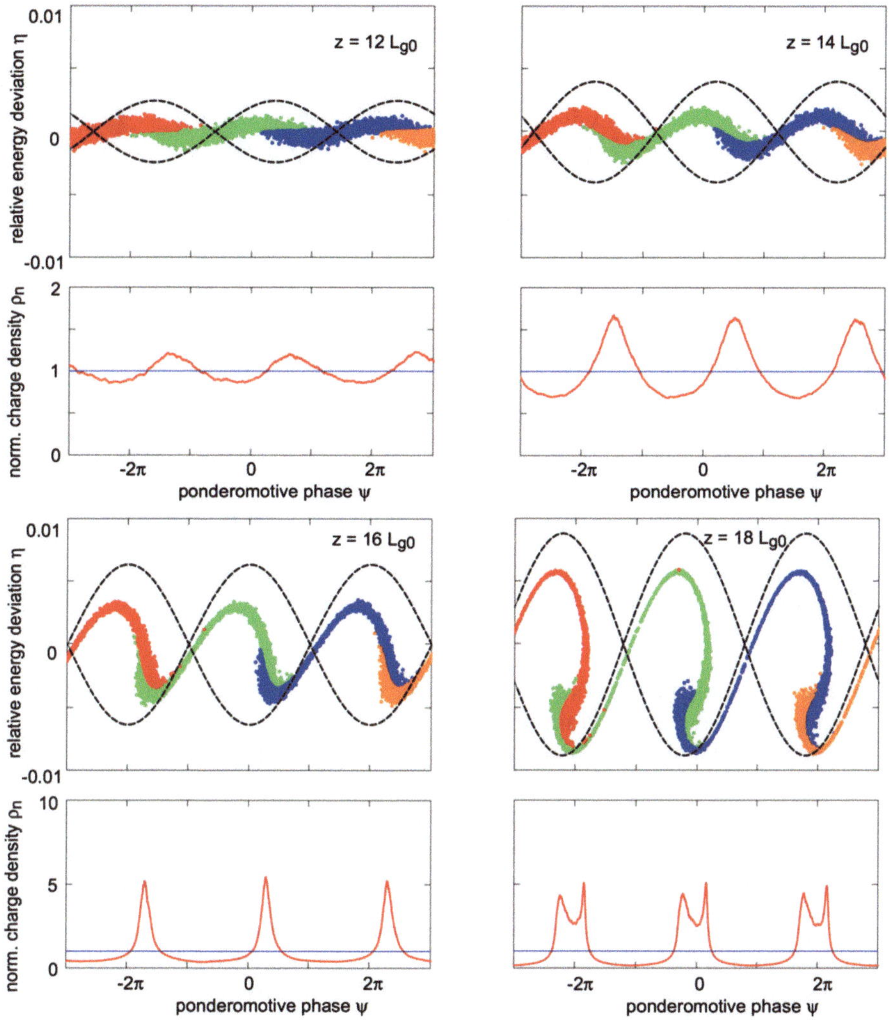

Fig. 5.9 Evolution of the microbunch structure. *Top left*: $z = 12\,L_{g0}$, *top right*: $z = 14\,L_{g0}$. *Bottom left*: $z = 16\,L_{g0}$, *bottom right*: $z = 18\,L_{g0}$. The upper subplots show the distribution of the particles in the (ψ, η) phase space. The separatrices are indicated by the *dashed curves*. The lower subplots show the normalized particle density as a function of ψ.

of a pronounced microbunch structure. The microstructure with the periodicity of the light wavelength λ_ℓ is fully developed after 16 power gain lengths.
We make several important observations:

(1) With increasing z the FEL buckets move towards smaller phases and the amplitude of the separatrix grows. The z dependence of the bucket-center phase will be determined in the next section.

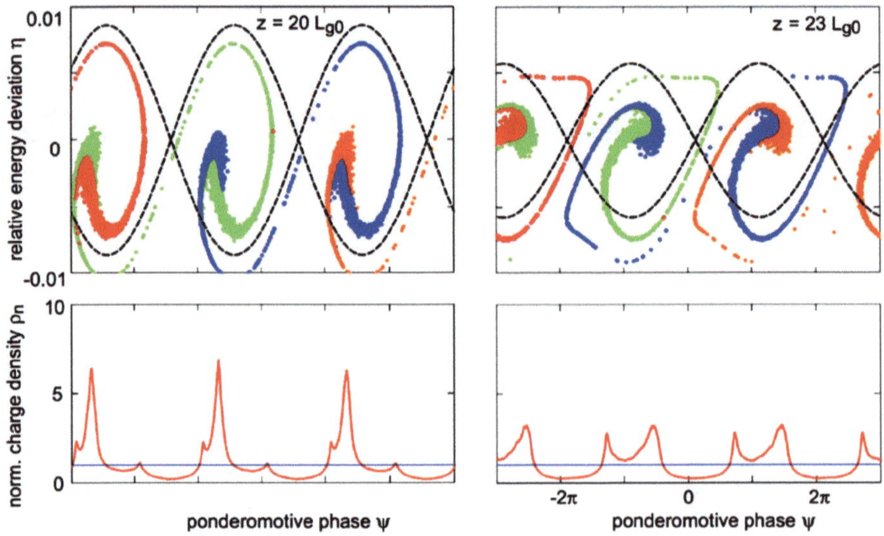

Fig. 5.10 Distortion of the phase space distribution and evolution of substructure of the microbunches deep in the saturation regime at 20 and 23 gain lengths.

(2) The energy modulation is almost harmonic up to 12–14 gain lengths but acquires anharmonic distortions for $z \geq 14\, L_{g0}$.
(3) Due to the bucket motion many particles do not stay in their original bucket but are captured by the next one.
(4) The normalized particle density gradually develops narrow peaks which are located in the right half of the respective FEL bucket, namely at the phase values where energy transfer from the electrons to the FEL wave will happen.

Deep in the saturation regime the phase space distributions acquire large distortions as shown in Fig. 5.10. Many particles move over into the left halves of the FEL buckets with the consequence that energy is withdrawn from the light wave. This is the reason for the reduction of the FEL power at about 20 gain lengths that is visible in Fig. 5.4. The microbunches develop a substructure. The energy spread of the beam increases considerably with increasing undulator length. The distortions of the particle distribution in phase space and the oscillatory behavior of the FEL gain curve in the saturation regime are well-known from the literature, see e.g. [1, 2].

5.6.2 Evolution of Phases in the FEL Gain Process

Phases of electric field and current density

To determine the phase evolution of the field $\tilde{E}_x(z)$ and of the current density $\tilde{j}_1(z)$ in the linear FEL regime, we express the slowly varying complex field amplitude as a superposition of the three eigenfunctions of the third-order equation

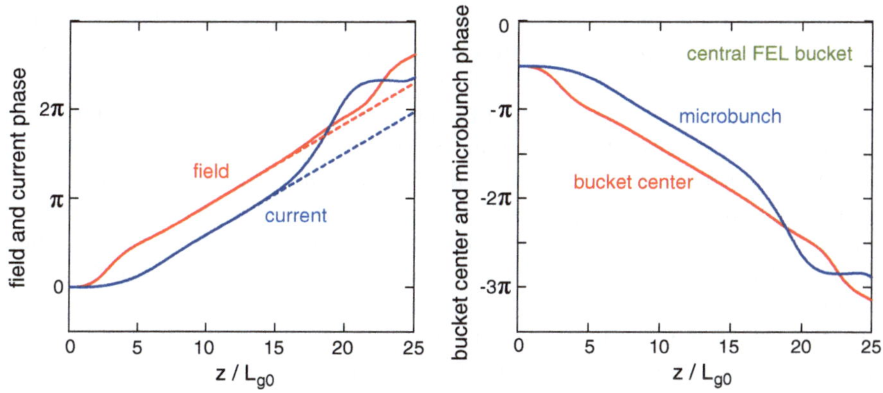

Fig. 5.11 *Left*: The phase $\varphi_E(z)$ of the complex field $\tilde{E}_x(z)$ (*solid red curve*) and the phase $\varphi_{j1}(z)$ of the modulated current density $\tilde{j}_1(z)$ (*solid blue curve*) as a function of z/L_{g0}. The phases have been computed using the coupled first-order equations. In the linear regime there is perfect agreement with the phases derived from the Eqs. (5.21) and (5.22). The dashed lines show the extrapolation of the linear theory into the nonlinear regime at $z > 17\,L_{g0}$. *Right*: The bucket-center phase $\psi_b(z)$ (*red curve*) and the microbunch phase $\psi_m(z)$ (*blue curve*) versus z/L_{g0}.

$$\tilde{E}_x(z) = \frac{E_{\rm in}}{3}\sum_{j=1}^{3}\exp(\alpha_j z) \equiv \left|\tilde{E}_x(z)\right| \exp(i\,\varphi_E(z))\,. \tag{5.21}$$

In the simplest case, $\eta = 0$ and $k_p = 0$, the eigenvalues α_j are given by Eq. (4.49). The phase $\varphi_E(z)$ is readily computed from (5.21). The phase of the modulated current density $\tilde{j}_1(z)$ can be related to the phase of $\tilde{E}_x(z)$ by means of Eq. (4.21)

$$\tilde{j}_1(z) = -\frac{4\gamma_r}{\mu_0 c\,\hat{K}}\,\tilde{E}'_x(z) = -\frac{4\gamma_r}{\mu_0 c\,\hat{K}}\sum_{j=1}^{3}\alpha_j\exp(\alpha_j z) \equiv \left|\tilde{j}_1(z)\right| \exp(i\,\varphi_{j1}(z))\,. \tag{5.22}$$

The z dependence of the two phases is depicted in the left part of Fig. 5.11. The phase of the field grows initially faster than that of the current. The phase difference $\Delta\varphi = [\varphi_E(z) - \varphi_E(0)] - [\varphi_{j1}(z) - \varphi_{j1}(0)]$ grows from $\Delta\varphi = 0$ at small z to $\Delta\varphi = \pi/3$ for $4L_{g0} \le z \le 17\,L_{g0}$. In the saturation regime the coupled first-order equations must be used. One finds a cross-over of the phases at about 20 gain lengths.

Bucket center phase and microbunch phase

In Sect. 4.2 we have discussed the origin of microbunching as being due to a modulation of the longitudinal velocity: particles losing energy to the light wave move on an undulator trajectory of larger amplitude and have a smaller average speed in z direction than particles gaining energy from the light wave. This reasoning is unable to predict the position of the microbunches inside the FEL buckets. Now we address this question which is of crucial importance for the functioning of the high-gain FEL:

why are the microbunches located in the right half of the respective FEL bucket where energy is transferred from the electron beam to the light wave, and why not in the left half where energy would be withdrawn from the light wave? The explanation is based on the observation that the FEL buckets move in phase space during the exponential gain process. The peak of the charge density modulation moves also, but at a slower rate, and the resulting phase difference turns out to be essential.

The coupled first-order equation (4.31b) can be utilized to determine the phase evolution of the bucket center. The FEL buckets are generated by the electric field $\tilde{E}_x$ of the light wave, and therefore only the first term in this equation has to be considered while the second term, containing the current $\tilde{j}_1(z)$, is neglected. We consider in the following the central FEL bucket and represent the bucket center by a test particle "b". Equation (4.31b) reads for this particle

$$\frac{d\eta_b}{dz} = -\frac{e}{m_e c^2 \gamma_r} \Re\left\{ \frac{\widehat{K}\tilde{E}_x(z)}{2\gamma_r} \exp[i\,\psi_b(z)] \right\} \propto \cos[\varphi_E(z) + \psi_b(z)]\,,$$

where $\varphi_E(z)$ is the z dependent phase of the complex field $\tilde{E}_x(z)$. The energy of particle "b" has to remain invariant if this particle is supposed to represent the bucket center for all $z \geq 0$. Hence we request $d\eta_b/dz = 0$ from which follows

$$\varphi_E(z) + \psi_b(z) = \text{const} = \varphi_E(0) + \psi_b(0)\,.$$

Since $\varphi_E(0) = 0$ and $\psi_b(0) = -\pi/2$ (central FEL bucket) we obtain a relation between the bucket-center phase and the phase of the complex field $\tilde{E}_x(z)$:

$$\psi_b(z) = -\pi/2 - \varphi_E(z)\,. \tag{5.23}$$

In a similar way, Eq. (4.31b) can be utilized to determine the phase evolution of the peak of the charge distribution, namely the center position of the "microbunch". In this case the $\tilde{E}_x(z)$ term has to be ignored and the $\tilde{j}_1(z)$ term is relevant because it describes the particle's interaction with the space charge field generated by the microbunch. From the condition $d\eta_c/dz = 0$ follows that the "microbunch phase" obeys the equation

$$\psi_m(z) = -\pi/2 - \varphi_{j1}(z)\,. \tag{5.24}$$

The sign reversal between $\psi_b(z)$ and $\varphi_E(z)$ and between $\psi_m(z)$ and $\varphi_{j1}(z)$ is very important. We have seen that the phases of the complex field $\tilde{E}_x(z)$ and the modulated current density $\tilde{j}_1(z)$ grow with increasing z. However, owing to the sign reversal, the bucket-center phase and the microbunch phase drop with increasing z, as depicted in the right part of Fig. 5.11. An immediate consequence is: the microbunch phase becomes larger than the bucket-center phase after a few gain lengths, in other words, the microbunches are located in the right half of the FEL bucket, as desired for FEL operation.

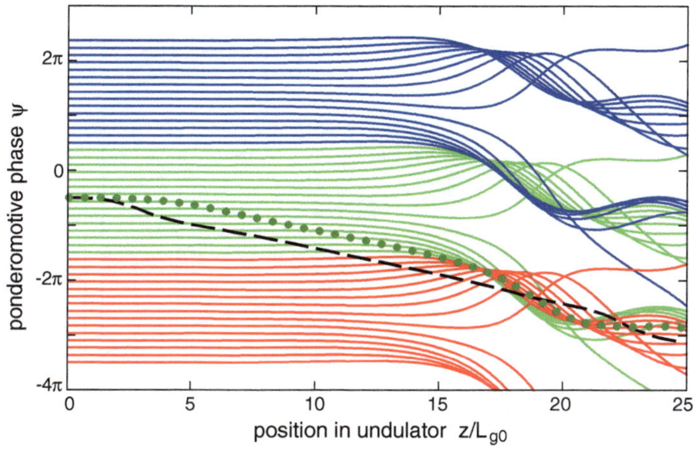

Fig. 5.12 Phase motion of selected particles in three adjacent FEL buckets during the exponential growth and saturation of the FEL power. The beam is mono-energetic and has an initial fractional energy deviation of $\eta = 0$. The *dashed curve* shows the variation of the FEL bucket center phase $\psi_b(z)$ according to (5.23). The *dotted curve* shows the motion of the microbunch phase $\psi_m(z)$ as given by Eq. (5.24).

Evolution of particle phases in the gain process

The phase evolution of a selected number of particles in three adjacent FEL buckets is depicted in Fig. 5.12. The ponderomotive phases $\psi_n = \psi_n(z)$ have been computed with the coupled equations (4.31) and are plotted as functions of the distance z traveled in the undulator.

The particles are initially on resonance ($W = W_r$, $\eta = 0$). In the range $0 \leq z \leq 10\,L_{g0}$ the particle phases $\psi_n(z)$ remain almost invariant. For $z > 13\,L_{g0}$ a dramatic variation of the particle phases is observed. They begin to concentrate in narrow bands which are fully developed at 18–20 gain lengths. These bands are the microbunches.

We look specifically at the central bucket which comprises initially the phase range $-3\pi/2 < \psi < \pi/2$. Bucket-center phase and microbunch phase are identical at the origin, $\psi_b(0) = \psi_m(0) = -\pi/2$. We see in Fig. 5.13 that the whole bucket moves in the negative ψ direction while the particle phases are almost stationary. About half of the particles remain in their bucket, and the other half is captured by the next bucket.

The microbunching process is thus associated with a mixing of neighboring buckets. Moreover it becomes evident that the microbunch is formed in the right half of the bucket. This is the explanation for the strong rise of the FEL power.

We emphasize again that the phase $\psi_b(z)$ of the bucket center as well as the phase $\psi_m(z)$ of the microbunch move with respect to the particles. In the exponential gain regime, $4\,L_{g0} \leq z \leq 17\,L_{g0}$, we find

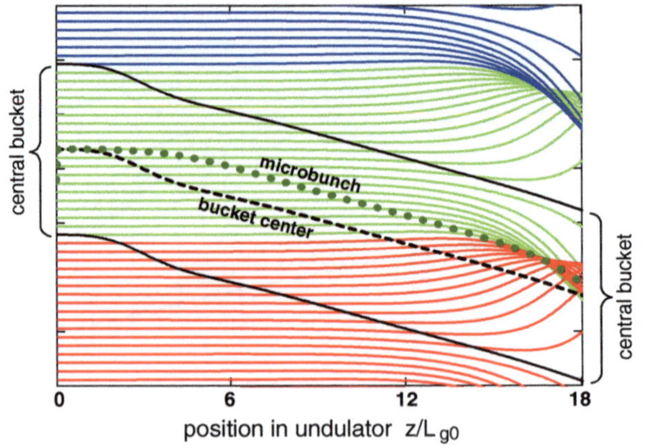

Fig. 5.13 Motion of the central FEL bucket for $0 \leq z \leq 18 L_{\mathrm{g}0}$.

$$\Delta\psi = \psi_m(z) - \psi_b(z) \approx \pi/3.$$

The peaks in the charge distribution are localized in the right half of the respective bucket, as required for energy transfer from the electron beam to the light wave. Above 20 gain lengths, however, the maximum of the charge density moves over to the left half of the bucket, corresponding to energy transfer from the light wave to the electron beam.

5.6.3 The Onset of Microbunching

It may appear strange at first sight that the particle phases are nearly invariant in the lethargy regime and in a large part of the exponential gain regime, while the charge density modulation is moving. This, however, is quite a common phenomenon. An example is a sound wave in air. The molecules carry out small longitudinal oscillations about stationary positions, but the density fluctuations propagate with the speed of sound.

To demonstrate explicitly that a density modulation evolves already quite early in the undulator we show in Fig. 5.14 the normalized charge density within the central FEL bucket at various longitudinal positions z. At the very beginning the maximum of the charge density is found at the bucket center at $\zeta = 0$, but already after two gain lengths the maximum begins to move towards positive ζ values. In the exponential gain regime ($4\,L_{\mathrm{g}0} \leq z \leq 17\,L_{\mathrm{g}0}$) the maximum is located in the right half of the FEL bucket at $\zeta \approx \lambda_\ell/6$, corresponding to $\Delta\psi \approx \pi/3$, see Figs. 5.9 and 5.14.

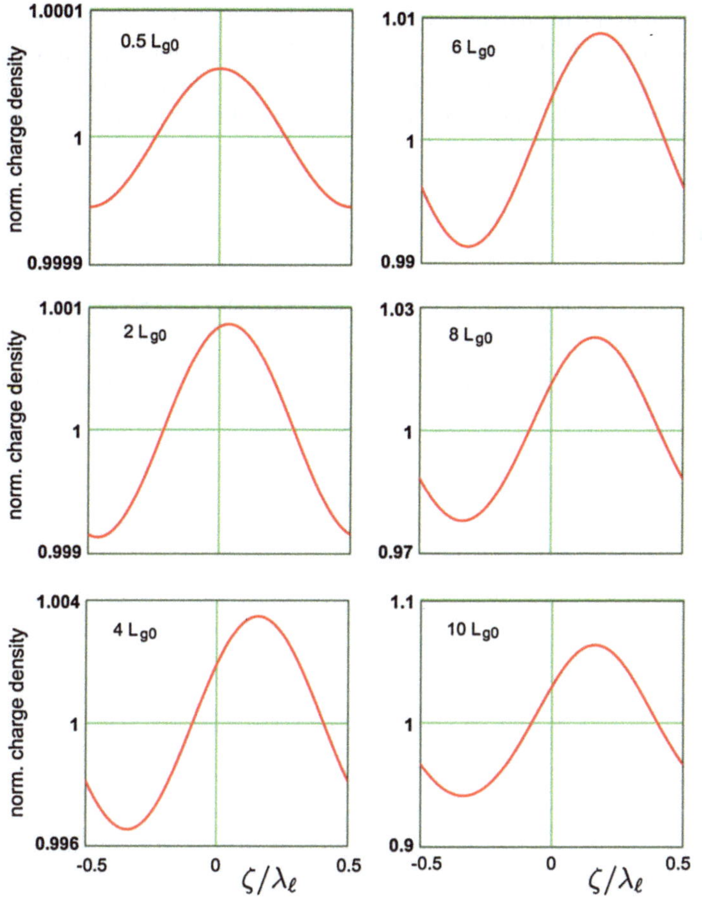

Fig. 5.14 The normalized charge density distribution within the central FEL bucket at $z = 0.5\,L_{g0}, 2\,L_{g0}, 4\,L_{g0}, 6\,L_{g0}, 8\,L_{g0}, 10\,L_{g0}$. Note the suppressed zero in the *vertical scales*. Only a narrow range around $\rho_n = 1$ is displayed.

5.6.4 Higher Harmonics in FEL Radiation

Figure 5.14 displays the normalized charge density distribution inside an FEL bucket for the range $0.5\,L_{g0} \leq z \leq 10\,L_{g0}$. A flat distribution ρ_0 is superimposed with a small, almost harmonic modulation. At $z \geq 12\,L_{g0}$ the picture changes dramatically. It is evident from Fig. 5.9 that the longitudinal charge distribution has still the periodicity of the fundamental light wavelength $\lambda_1 \equiv \lambda_\ell$ but is no longer a simple harmonic distribution. Its Fourier decomposition will therefore contain higher harmonics. These higher harmonics, however, will not be amplified in our simplified one-dimensional FEL model because we have averaged over the oscillatory part in the longitudinal velocity, compare Eq. (4.6). It is possible to generalize the coupled first-order equations in such a way that the odd higher harmonics are incorporated. In three-dimensional FEL codes such as GENESIS the longitudinal oscillation is properly accounted for, and then higher harmonics can be computed as well. There is ample experimental evidence for the existence of higher harmonics, see Chap. 7.

References

1. J.B. Murphy, C. Pellegrini, Introduction to the physics of the free electron laser, in *Laser Handbook*, ed. by W.B. Colson, C. Pellegrini, A. Renieri, Vol. 6 (North Holland, Amsterdam, 1990), p. 11
2. W.B. Colson, Classical free electron laser theory, *Laser Handbook* (North Holland, Amsterdam, 1990), p. 115

Chapter 6
Energy Spread, Space Charge and 3D Effects

The realistic description of FELs operating in the high-gain regime has to be based on a three-dimensional theory. The dependencies of the electron beam current density and of the light wave on the transverse coordinates x and y must be taken into account. Betatron oscillations of the electrons and diffraction of the light wave play an important role. Moreover, energy spread in the electron beam and the longitudinal slippage of the FEL pulse with respect to the short electron bunch must be considered (in the 1D theory as described in Chaps. 4 and 5 the bunch is treated as being infinitely long, which is far from reality). No analytical methods are known to carry through the full 3D theory and one has to resort to numerical simulation codes such as GINGER [1], GENESIS [2] or FAST [3] which have been developed to a high degree of sophistication. The full three-dimensional treatment of the FEL is beyond the scope of this introductory book, the interested reader is referred to the monograph *The Physics of Free Electron Lasers* by Saldin, Schneidmiller and Yurkov [4] and to the review article *Review of x-ray free-electron laser theory* by Huang and Kim [5] and the literature quoted therein.

In this chapter we consider first the effects that are already present in the 1D theory: energy detuning, beam energy spread and space charge forces, and we analyze their impact on the FEL gain length. Then 3D effects are treated in a somewhat qualitative manner in order to explain the influence of betatron oscillations and optical diffraction on the FEL performance. A simplified three-dimensional model is applied to determine numerically the FEL growth rate parameter and the gain length for a cylindrical electron beam of uniform charge density. In Sect. 6.4, X-ray FEL parameter studies of other authors are presented which were derived from numerical simulations. Finally, slippage effects in short electron bunches are discussed.

P. Schmüser et al., *Free-Electron Lasers in the Ultraviolet and X-Ray Regime*, Springer Tracts in Modern Physics 258, DOI: 10.1007/978-3-319-04081-3_6,

6.1 Increase of Gain Length by Energy Spread and Space Charge

In Sect. 4.8 the FEL power gain length has been defined in the 1D approximation and for a mono-energetic beam that is on resonance. According to Eq. (4.50) the power gain length is related to the real part of the first eigenvalue of the third-order equation:

$$P(z) \propto \exp(2\,\Re\{\alpha_1\}\,z) \equiv \exp(z/L_{g0})\,. \tag{6.1}$$

We infer that $2\,\Re\{\alpha_1\}$ is the inverse gain length. This insight allows us to compute the gain length in more general cases.

6.1.1 Energy Detuning

We consider first energy detuning, $W \neq W_r, \eta \neq 0$, but we keep $\sigma_\eta = 0$ and $k_p = 0$. It is shown in Sect. 10.4 how the eigenvalues $\alpha_j = \alpha_j(\eta)$ of the third-order differential equation can be determined for the case $\eta \neq 0$. The real parts of the eigenvalues are depicted in Fig. 6.1.

The real part of the first eigenvalue determines again the growth rate but it is now a function of the relative energy deviation η. According to Eq. (6.1) it appears reasonable to define the inverse gain length by the maximum value of $2\,\Re\{\alpha_1(\eta)\}$. So the gain length is given by the expression

$$\boxed{L_g = \frac{1}{\max\,[2\,\Re\{\alpha_1(\eta)\}]}}\,. \tag{6.2}$$

In this special case of a beam without energy spread and space charge we obtain $L_g = L_{g0}$. For the investigation of the general case it is convenient to define a dimensionless "growth rate function" by

$$f_{\mathrm{gr}}(\eta) = 2\,\Re\{\alpha_1(\eta)\}\,L_{g0}\,. \tag{6.3}$$

The maximum value of the growth rate function is equal to the ratio L_{g0}/L_g:

$$\max\left[f_{\mathrm{gr}}(\eta)\right] = \max\left[2\,\Re\{\alpha_1(\eta)\}\,L_{g0}\right] = \frac{L_{g0}}{L_g}\,.$$

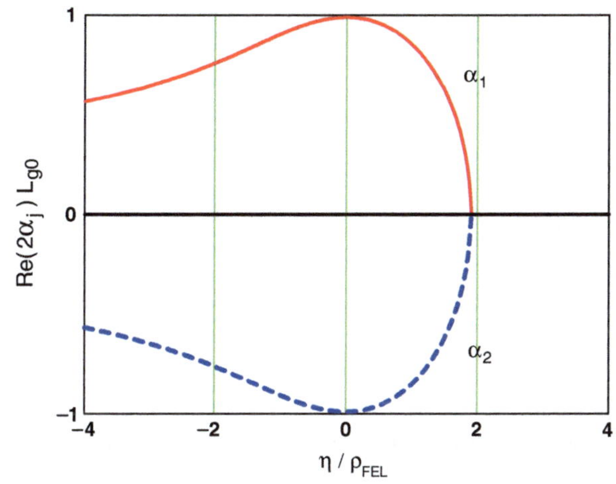

Fig. 6.1 The real part of the first and the second eigenvalue, multiplied with $2L_{g0}$, is plotted as a function of $\eta/\rho_{\rm FEL}$, the relative energy deviation divided by the FEL parameter. Note that $\Re(\alpha_1)$ (*continuous red curve*) is positive, corresponding to exponential growth of the eigenfunction $V_1(z) = \exp(\alpha_1 z)$. However, the real part vanishes above $\eta \approx 1.88\,\rho_{\rm FEL}$ which means that the exponential growth stops if the electron energy W exceeds the resonant energy W_r by more than $\Delta W = 1.88\,\rho_{\rm FEL}\,W_r$. The real part of α_2 (*dashed blue curve*) is always negative, hence the eigenfunction $V_2(z)$ drops exponentially. Note that $\Re(\alpha_3) \equiv 0$, so $V_3(z)$ oscillates along the undulator axis.

6.1.2 Energy Spread and Space Charge

Now we drop the assumption of a mono-energetic beam and admit energy spread ($\sigma_\eta > 0$) and space charge ($k_p > 0$). The mean beam energy is called W_0, it may be different from the resonance energy ($W_0 \neq W_r$, $\eta_0 \neq 0$). The third-order equation is not suitable for a beam with energy spread but the integro-differential Eq. (10.51) can be used. The growth rate function $f_{\rm gr}(\eta_0) = 2\,\Re\{\alpha_1(\eta_0)\}\,L_{g0}$ depends on k_p and σ_η. Numerical results are shown in Fig. 6.2.

The impact of space charge forces on the gain length is often rather small. The gain length increases by 10 % if the space charge parameter has the value $k_p = 0.5\,\Gamma$ and by 30 % for $k_p = 1.0\,\Gamma$. The space charge forces can thus be neglected if

$$k_p \ll \Gamma\ . \tag{6.4}$$

At FLASH, one finds $k_p \approx 0.2\,\Gamma$, and the increase of the gain length is less than 1 %.

The influence of beam energy spread is bigger. For $\sigma_\eta = 0$, the growth rate function has a maximum value of 1 which means that the power gain length L_g is indeed identical to L_{g0} (this case has already been discussed in the previous section). However, for an energy spread of $\sigma_\eta = 0.5\,\rho_{\rm FEL}$, the maximum drops to 0.8, implying that the gain length is 25 % larger than the ideal 1D gain length, $L_g = 1.25\,L_{g0}$. At $\sigma_\eta = \rho_{\rm FEL}$ the gain length increases by more than a factor of two. Consequently,

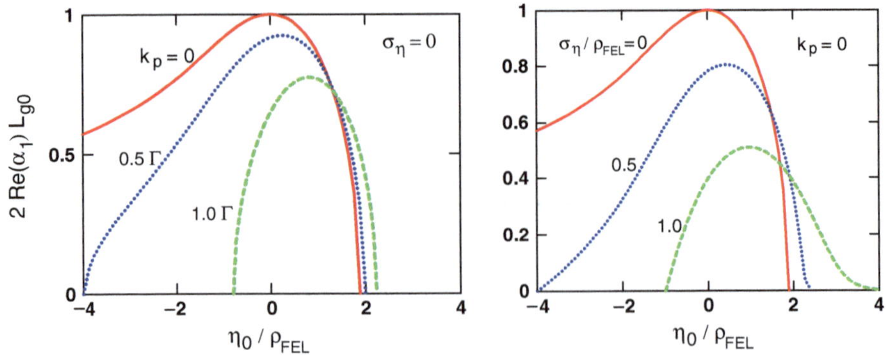

Fig. 6.2 *Left*: The growth rate function $f_{gr}(\eta_0) = 2\,\Re\{\alpha_1(\eta_0)\}\,L_{g0}$ is plotted versus $\eta_0/\rho_{\rm FEL}$ for different values of the space charge parameter k_p. *Solid red curve*: $k_p = 0$, *dotted blue curve*: $k_p = 0.5\,\Gamma$, *dashed green curve*: $k_p = 1.0\,\Gamma$. The energy spread is put to zero. *Right*: The growth rate function $f_{gr}(\eta_0) = 2\,\Re\{\alpha_1(\eta_0)\}\,L_{g0}$ is plotted versus $\eta_0/\rho_{\rm FEL}$ for different values of the relative beam energy spread $\sigma_\eta = \sigma_W/W_r$. *Solid red curve*: $\sigma_\eta = 0$, *dotted blue curve*: $\sigma_\eta = 0.5\,\rho_{\rm FEL}$, *dashed green curve*: $\sigma_\eta = 1.0\,\rho_{\rm FEL}$. Here the space charge parameter is set to zero.

the energy spread of the electron beam must be sufficiently small because only the particles inside a narrow energy window contribute constructively to the FEL gain process. A reasonable upper limit for the tolerable rms energy spread is

$$\frac{\sigma_W}{W_r} \equiv \sigma_\eta < 0.5\,\rho_{\rm FEL}\,. \tag{6.5}$$

The beam energy spread at FLASH is $\sigma_\eta \approx 0.7 \cdot 10^{-3} \approx 0.2\,\rho_{\rm FEL}$, which means that the increase in gain length due to energy spread is less than 10 %. Note, however, that the achievement of a low energy spread is a serious technical challenge for FELs operating in the X-ray regime where the FEL parameter becomes quite small, for instance $\rho_{\rm FEL} = 5 \cdot 10^{-4}$ in LCLS (see Chap. 9).

It is important to realize that any kind of energy variation inside the electron bunch, and not only the energy spread, will have an impact on the FEL gain.

6.2 Increase of Gain Length by 3D Effects

Up to now we have neglected the transverse dimensions of the electron beam. In Sect. 10.4 the three eigenvalues α_j of the third-order equation are determined for a cylindrical bunch with radius r_b and homogeneous charge density. The amplitude of the electric field can be expressed by the Bessel function J_0 for $r < r_b$ and by the modified Bessel function K_0 for $r > r_b$. The boundary conditions at $r = r_b$ lead to a determinant equation which is solved by numerical iteration and yields the eigenvalues α_j.

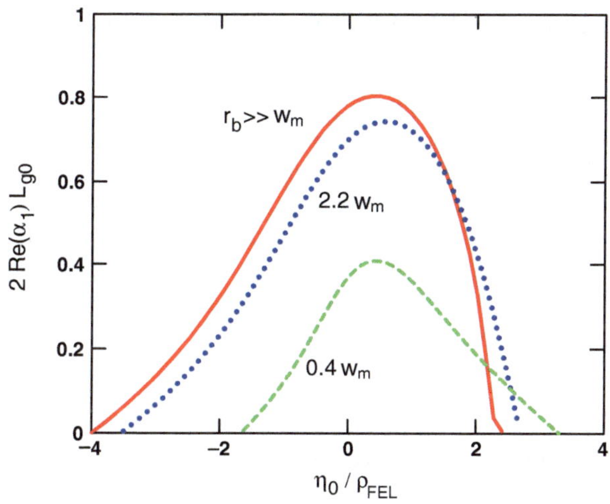

Fig. 6.3 The growth rate function $f_{\mathrm{gr}}(\eta_0) = 2\,\Re\{\alpha_1(\eta_0)\}\,L_{\mathrm{g0}}$ plotted vs. $\eta_0/\rho_{\mathrm{FEL}}$ for a cylindrical beam with energy spread ($\sigma_\eta = 0.5\,\rho_{\mathrm{FEL}}$). *Continuous red curve*: wide beam with a radius $r_b \gg w_m = \sqrt{L_{\mathrm{g0}}\lambda_\ell}$, *dotted blue curve*: narrow beam $r_b = 2.2\,w_m$, *dashed green curve*: very narrow beam, $r_b = 0.4\,w_m$.

Is it possible to establish a criterion under which circumstances the 1D theory is sufficient or if a more complicated 3D treatment is needed? It turns out useful to define a scale parameter for the radial width of the electron beam, namely the geometric mean of power gain length and optical wavelength:

$$w_m = \sqrt{L_{\mathrm{g0}}\,\lambda_\ell}\ . \tag{6.6}$$

The growth rate function $f_{\mathrm{gr}}(\eta_0) = 2\,\Re\{\alpha_1(\eta_0)\}\,L_{\mathrm{g0}}$ is shown in Fig. 6.3 for a cylindrical beam of constant charge density. Three different beam radii are considered: $r_b \gg w_m$, $r_b = 2.2\,w_m$, and $r_b = 0.4\,w_m$. We present here the computation for a beam with an energy spread of $\sigma_\eta = 0.5\,\rho_{\mathrm{FEL}}$. According to Fig. 6.2 the 1D gain length is then $L_g^{\mathrm{1D}} \approx 1.25\,L_{\mathrm{g0}}$. From the curves in Fig. 6.3 it is obvious that in case of a "thick" beam the gain length retains its 1D value of about $1.25\,L_{\mathrm{g0}}$ while for a very "thin" electron beam ($r_b < \sqrt{L_{\mathrm{g0}}\,\lambda_\ell}$) a considerable further increase in gain length has to be expected, for example by a factor of 2.5 for $r_b = 0.4\sqrt{L_{\mathrm{g0}}\,\lambda_\ell}$. Hence the one-dimensional FEL theory is adequate if the criterion

$$r_b \gg \sqrt{L_{\mathrm{g0}}\,\lambda_\ell} \tag{6.7}$$

is satisfied (see also Sect. 10.4.4), but a three-dimensional treatment is needed for beams of small radius. The 1D theory is usually insufficient for a quantitative description of FELs in the UV and X-ray regime.

6.3 Overlap Between Electron and Photon Beam

The amplification process in the high-gain FEL depends critically on a good transverse overlap between electron and photon beam. Intuitively, both beams should have the same cross section to ensure that the energy transfer is optimized all along the undulator. Considering the fact that the typical diameter of a GeV electron beam is below 100 μm, the undulator must be manufactured and aligned with great precision such that the deviation of the electrons from the design orbit is not larger than 10 μm over several gain lengths. This puts stringent requirements on the field quality and mechanical accuracy of the undulator.

6.3.1 Electron Beam Focusing and Emittance

Transverse focusing is mandatory in the long undulator structures of high-gain FELs. The so-called "natural focusing" in an undulator is analogous to the "weak focusing" in circular accelerators [6]. A horizontally deflecting dipole magnet of rectangular shape exerts a weak vertical focusing which is caused by end field effects. A planar undulator is a sequence of short rectangular dipole magnets and provides thus weak focusing in vertical direction. This "natural focusing" has two disadvantages: (1) it acts only in the vertical plane, while the horizontal motion remains unfocused, and (2) it is too weak for short-wavelength FELs. To avoid the first disadvantage, Scharlemann [7] proposed to shape the pole faces to be parabolic. In this way, the natural focusing can be distributed equally among both transverse directions. This solution has not been widely pursued because of the mechanical complications and the inability to adjust the beta function to the needs of the FEL. The generally adopted solution is to apply the principle of "strong focusing" [6], known from synchrotrons, by augmenting the undulator system with a periodic lattice of quadrupole lenses of alternating polarity, a so-called FODO lattice where F denotes a focusing quadrupole, D a defocusing quadrupole, and O a drift space or a weakly focusing element such as a bending magnet or an undulator. This FODO lattice can be realized either by superimposing quadrupole fields inside the undulator itself or, more conveniently, by placing electromagnetic quadrupoles between the segments of a long undulator structure. Since these segments are typically 5 m long, the period of the FODO lattice is in the order of 10 m and can be easily tuned to yield beta functions in the range from 5 to 30 m.

The rms horizontal width of the electron beam at the position z is computed from the equation

$$\sigma_x(z) = \sqrt{\langle x^2(z)\rangle - \langle x(z)\rangle^2} = \sqrt{\varepsilon_x \beta_x(z)} \quad \text{with} \quad x^2(z) = [x(z)]^2 \,. \tag{6.8}$$

Here $\beta_x(z)$ is the horizontal beta function inside the undulator and ε_x is the horizontal emittance. A similar equation holds for the rms vertical width

$$\sigma_y(z) = \sqrt{\langle y^2(z)\rangle - \langle y(z)\rangle^2} = \sqrt{\varepsilon_y \beta_y(z)}\ . \tag{6.9}$$

The concept of the beta function is explained in any textbook on accelerator physics, see for example Ref. [6]. The brackets $\langle\ \rangle$ in Eqs. (6.8) and (6.9) indicate an averaging over the betatron oscillation amplitudes and phases in the beam. The mean squared slopes of the particle trajectory are given by [6]

$$\left\langle x'^2(z)\right\rangle = \frac{1+(\beta_x'(z)/2)^2}{\beta_x(z)}\,\varepsilon_x \quad \text{and} \quad \left\langle y'^2(z)\right\rangle = \frac{1+(\beta_y'(z)/2)^2}{\beta_y(z)}\,\varepsilon_y\ , \tag{6.10}$$

where $x' = dx/dz$ and $y' = dy/dz$ are the slopes of the particle trajectory in the horizontal and vertical plane, respectively.

The emittance is, roughly speaking, the product of beam size and beam divergence. The statistical definition is

$$\boxed{\varepsilon_x = \sqrt{\langle x^2\rangle \cdot \langle x'^2\rangle - \langle x\cdot x'\rangle^2}\,,\quad \varepsilon_y = \sqrt{\langle y^2\rangle \cdot \langle y'^2\rangle - \langle y\cdot y'\rangle^2}} \tag{6.11}$$

with the subsidiary condition that the beam is centered in phase space, i.e.

$$\langle x\rangle = \langle x'\rangle = 0\,,\quad \langle y\rangle = \langle y'\rangle = 0\ .$$

In the following rough estimations one can ignore the z dependence of the beta function. Hence we drop the derivatives β_x' and β_y' in (6.10) and average the beta function over the longitudinal coordinate z, assuming

$$\overline{\beta_x(z)} = \overline{\beta_y(z)} \equiv \beta_{\rm av}\ . \tag{6.12}$$

Furthermore, we assume a round beam ($\varepsilon_x = \varepsilon_y \equiv \varepsilon$). The emittance shrinks like $1/\gamma$ with increasing electron energy [6]. This is easy to understand: during acceleration the longitudinal electron momentum grows like γ while the transverse momentum remains invariant, hence the beam divergence shrinks. At relativistic electron accelerators is convention to define the normalized emittance by

$$\boxed{\varepsilon_n = \gamma\,\varepsilon\ .} \tag{6.13}$$

Ideally the normalized emittance ε_n should remain constant all along the accelerator. In practice, space charge forces, radiation effects and wake fields often lead to an emittance growth and a dilution of the particle density in the beam.

Besides determining the beam size and divergence, the emittance has an additional influence on the lasing process. The horizontal and vertical betatron oscillations introduce additional transverse velocity components. The average longitudinal speed of a particle carrying out betatron oscillations in the undulator is lower than the speed $\bar{v}_z$ in Eq. (2.11)

$$\bar{v}_z^\beta = \sqrt{\bar{v}_z^2 - \langle v_x^2 \rangle - \left\langle v_y^2 \right\rangle} \approx \bar{v}_z - \frac{\langle v_x^2 \rangle}{2c} - \frac{\left\langle v_y^2 \right\rangle}{2c} \,,$$

where the superscript "β" indicates the presence of a betatron oscillation. Using Eq. (6.10) with $\beta_x' = 0$ and $\beta_y' = 0$ we get

$$\left\langle v_x^2 \right\rangle \approx c^2 \left\langle x'^2 \right\rangle = c^2 \frac{\varepsilon}{\beta_{\mathrm{av}}} \,, \quad \left\langle v_y^2 \right\rangle \approx c^2 \frac{\varepsilon}{\beta_{\mathrm{av}}} \,.$$

The average longitudinal speed in a beam with betatron oscillations is therefore given by

$$\left\langle \bar{v}_z^\beta \right\rangle \approx \left(1 - \frac{1}{2\gamma^2}\left(1 + \frac{K^2}{2}\right)\right) c - \frac{\varepsilon}{\beta_{\mathrm{av}}} c \,. \tag{6.14}$$

Note that three different averaging procedures are applied in this expression: (1) the "bar" in $\bar{v}_z$ denotes the mean longitudinal speed of an electron moving through the undulator on its sinusoidal trajectory, but without betatron oscillation, (2) the brackets in $\left\langle \bar{v}_z^\beta \right\rangle$ indicate the averaging over the various betatron amplitudes in a beam, and (3) β_{av} is the longitudinal average (6.12) of the beta function.

Concerning the FEL synchronization process, the reduction of the longitudinal speed is equivalent to a reduction of the mean electron beam energy. The equivalent reduction of the Lorentz factor is computed as

$$(\delta\gamma)_{\mathrm{eq}} = \frac{d\gamma}{d\bar{v}_z} \delta\bar{v}_z \approx \frac{\gamma^3 \delta\bar{v}_z}{c} = -\frac{\gamma^3 \varepsilon}{\beta_{\mathrm{av}}} \,.$$

Owing to the fact that the particles in the beam have all different betatron oscillation amplitudes one obtains in fact not only a reduction in the longitudinal speed but in addition a smear which is equivalent to an energy spread of the incident beam:

$$\left(\sigma_\eta\right)_{\mathrm{eq}} = \frac{\left(\sigma_\gamma\right)_{\mathrm{eq}}}{\gamma_r} \approx \frac{\gamma_r^2 \varepsilon}{\beta_{\mathrm{av}}} \,.$$

The energy spread of the incident beam and the equivalent energy spread caused by betatron oscillations have to be added in quadrature as they stem from statistically independent sources. Requesting that the total energy spread stays below the tolerable value, which according to the inequality (6.5) is about $\rho_{\mathrm{FEL}}/2$, we derive an upper limit for the beam emittance:

$$\varepsilon < \frac{\beta_{\mathrm{av}}}{2\sqrt{2}\,\gamma_r^2} \rho_{\mathrm{FEL}} \,. \tag{6.15}$$

The magnitude of the transverse velocity can in principle be reduced by choosing a sufficiently large beta function. On the other hand, a larger beta function yields a

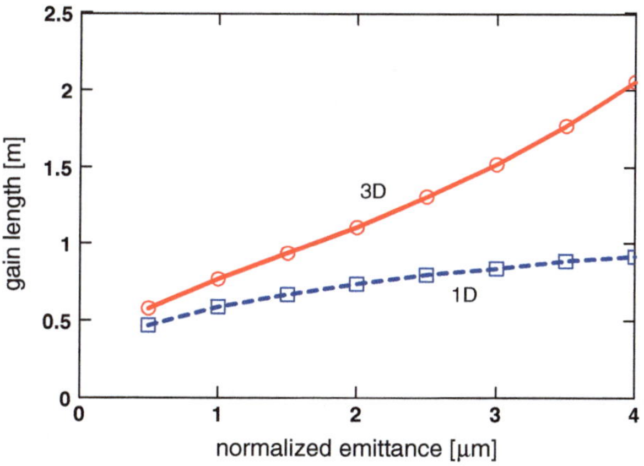

Fig. 6.4 The computed 3D power gain length L_g (*solid red curve*) and the 1D power gain length L_{g0} (*dashed blue curve*) as a function of the normalized emittance. The calculations are done for an energy of 1 GeV in FLASH and are based on Eqs. (4.51) and (6.22) with the following parameters: peak current $I_{\text{peak}} = 2500\,\text{A}$, average beta function $\beta_{\text{av}} = 5\,\text{m}$, energy spread $\sigma_\eta = 0.5\,\rho_{\text{FEL}}$.

lower electron density and a longer gain length, so there is an optimum in between. An estimate on this optimum beta function will be given below in Eq. (6.26). In the X-ray regime the optimum beta function is in the 10–30 m range.

A quite general criterion for the maximum tolerable emittance of the electron beam in a free-electron laser can be derived as follows. In Sect. 10.3 we show that one can define an emittance for a Gaussian laser beam as the product of rms beam width and divergence:

$$\varepsilon_{\text{laser}} = \sigma_x\,\sigma_\theta = \frac{\lambda}{4\pi}\,. \tag{6.16}$$

It is meaningful to request that the electron beam emittance does not exceed the light beam emittance because a broader electron beam would in turn generate a broader FEL beam. This leads to the emittance criterion

$$\boxed{\varepsilon \le \frac{\lambda_\ell}{4\pi}} \tag{6.17}$$

which is a very demanding requirement on the quality of the electron beam driving an X-ray FEL and cannot be fully satisfied in practice. In fact, the upper limit (6.17) is a bit too restrictive, a factor of 2–3 larger emittance is still sufficient for obtaining high FEL gain.

To illustrate the importance of a small emittance we show in Fig. 6.4 the gain length as a function of emittance for FLASH (see Chap. 8) at an energy of 1 GeV. The three-dimensional gain length has been computed with the Xie-formula (6.22) discussed below. One can see that the 3D power gain length L_g is rather close to the 1D gain length L_{g0} as long as the emittance criterion (6.17) is fulfilled. For FLASH operating at 1 GeV this corresponds to a normalized emittance $\varepsilon_n \leq 1\,\mu$m. If the criterion is violated the 3D gain length grows rapidly with increasing beam emittance while the 1D gain length exhibits only a moderate growth, $L_{g0} \propto \varepsilon_n^{1/3}$. Reducing the beam emittance is hence of utmost importance.

6.3.2 Optical Diffraction and Gain Guiding

Ideally the photon beam should have the same transverse size as the electron beam. However, like any electromagnetic wave, the FEL wave in the undulator undergoes optical diffraction. Since FEL radiation has a lot of similarity with optical laser beams we use here the Gaussian beam description, see Sect. 10.3. The Rayleigh length z_R is defined as the distance over which the beam cross section grows by a factor of 2 from its minimum value at a beam waist. It is related to the beam waist radius w_0 by

$$z_R = \frac{\pi\, w_0^2}{\lambda_\ell} \,. \tag{6.18}$$

A typical number is $z_R = 1$ m for $w_0 = 100\,\mu$m and $\lambda_\ell = 30$ nm. Moving away from the waist the beam radius grows as

$$w(z) = w_0 \sqrt{1 + \left(\frac{z - z_0}{z_R}\right)^2} \,. \tag{6.19}$$

Note that in Gaussian laser beam optics it is convention to define the radial width by the condition that the intensity of a TEM_{00} beam drops to $1/e^2 = 0.135$ of its value at $r = 0$ (the electric field drops to $1/e$).

The diffractive widening of a seed laser beam is not affected by the presence of the electron beam. It is therefore important to realize a sufficiently long Rayleigh length of this beam to obtain efficient seeding in the lethargy regime. Once the exponential FEL gain process has started, the additionally created field depends on the transverse dimensions of the electron beam and becomes decoupled from the seed beam.

The FEL beam will also be subject to diffraction, and the resulting widening could readily spoil the good overlap with the electron beam and reduce the energy transfer from the electrons to the light wave. Fortunately there exists an effect counteracting the widening of the FEL beam which is called gain guiding.[1] Gain guiding can be understood as follows. We consider an observation point z_0 in the exponential gain region. Most of the FEL intensity at this point has been produced in the last two or three gain lengths upstream of z_0, and the width of this newly generated radiation is determined by the electron beam width. The more distant contributions are widened by diffraction, however they play a minor role because they are much smaller in

[1] In the FEL literature, another effect named refractive or optical guiding has been discussed, see e.g. [8, 9].

amplitude. The overall result is an exponential growth of the central part of the light wave, and this part will retain its narrow width. Nevertheless, diffraction losses will occur. Three-dimensional numerical simulations show that some field energy evades radially from the light beam [10].

Another beneficial effect of gain guiding is the ability of the FEL beam to follow slow, "adiabatic" deviations of the electron beam away from its nominal orbit, which might be caused by spurious magnetic background fields. This guiding is important in the very long undulator magnets of an X-ray FEL.

To provide efficient gain guiding the FEL amplification has to be large enough so that the growth of the light intensity near the optical axis overcompensates the losses by diffraction. As a rule of thumb, the Rayleigh length has to be twice the power gain length or larger:

$$z_R \geq 2L_g\,. \tag{6.20}$$

This condition is fulfilled in practice. For example, in the soft X-ray FEL FLASH (see Chaps. 7, 8) the power gain length at $\lambda_\ell = 13\,\text{nm}$ amounts to $L_g = 1.25\,\text{m}$ and the Rayleigh length is $z_R = 3.6\,\text{m}$. In the hard X-ray FEL LCLS (see Chap. 9) the situation is even more favourable: the measured power gain length at $\lambda_\ell = 1.5\,\text{Å}$ is 3.5 m while the Rayleigh length is far longer and in the 50 m range.

6.4 Parametrization of Gain Length in an X-Ray FEL

We have seen above that the power gain length computed in the full three-dimensional theory will be larger than the 1D gain length of a mono-energetic beam. Many effects such as electron beam energy spread and emittance, space charge, finite bunch length and radiation diffraction play a role. For simplicity a round electron beam is considered with

$$\beta_x = \beta_y = \beta_{\text{av}}\,, \quad \varepsilon_x = \varepsilon_y = \varepsilon \quad \text{and} \quad \sigma_r = \sqrt{\varepsilon\,\beta_{\text{av}}}\,.$$

Three dimensionless parameters are useful in the characterization of a short-wavelength FEL. Following [11] we write them in the form

$$X_\gamma = \frac{L_{g0} 4\pi\sigma_\eta}{\lambda_u} \quad \text{energy spread parameter}\,, \tag{6.21a}$$

$$X_d = \frac{L_{g0}\lambda_\ell}{4\pi\sigma_r^2} \quad \text{diffraction parameter}\,, \tag{6.21b}$$

$$X_\varepsilon = \frac{L_{g0} 4\pi\varepsilon}{\beta_{\text{av}}\lambda_\ell} \quad \text{angular spread parameter}\,. \tag{6.21c}$$

From our previous considerations we can deduce that all these parameters should be less than 1. If we insert the FEL parameter (4.47) into the inequality (6.5) we obtain $X_\gamma < 1/\sqrt{3}$. The diffraction parameter is $X_d = L_{g0}/z_R \le 1/2$ according to Eq. (6.20). Finally, using Eqs. (4.47) and (6.15), the angular spread parameter can be written as

$$X_\varepsilon = \frac{\lambda_u \varepsilon}{\sqrt{3}\,\langle\beta_{\text{av}}\rangle\,\lambda_\ell\,\rho_{\text{FEL}}} < \frac{\lambda_u}{2\sqrt{2}\gamma_r^2}\,\frac{1}{\sqrt{3}\lambda_\ell} < \frac{1}{\sqrt{6}},$$

where we have used $\lambda_u/(2\gamma_r^2) < \lambda_\ell$.

M. Xie [11] has expressed the 3D gain length of an X-ray FEL in the form

$$L_g = L_{g0}(1 + \Lambda), \tag{6.22}$$

where $L_{g0} = \lambda_u/(4\pi\sqrt{3}\rho_{\text{FEL}})$ is the 1D gain length defined in Eq. (4.51). Based on three-dimensional numerical studies Xie obtained a parametrization of the correction term Λ in the form

$$\begin{aligned}\Lambda = a_1 X_d^{a_2} + a_3 X_\varepsilon^{a_4} + a_5 X_\gamma^{a_6} + a_7 X_\varepsilon^{a_8} X_\gamma^{a_9} + a_{10} X_d^{a_{11}} X_\gamma^{a_{12}} \\ + a_{13} X_d^{a_{14}} X_\varepsilon^{a_{15}} + a_{16} X_d^{a_{17}} X_\varepsilon^{a_{18}} X_\gamma^{a_{19}}\end{aligned} \tag{6.23}$$

with the following coefficients

$$\begin{aligned}&a_1 = 0.45,\ \ a_2 = 0.57,\ \ a_3 = 0.55,\ \ a_4 = 1.6,\ \ a_5 = 3.0,\\ &a_6 = 2.0,\ \ a_7 = 0.35,\ \ a_8 = 2.9,\ \ a_9 = 2.4,\ \ a_{10} = 51,\\ &a_{11} = 0.95,\ \ a_{12} = 3.0,\ \ a_{13} = 5.4,\ \ a_{14} = 0.7,\ \ a_{15} = 1.9,\\ &a_{16} = 1140,\ \ a_{17} = 2.2,\ \ a_{18} = 2.9,\ \ a_{19} = 3.2\,.\end{aligned}$$

Formula (6.23) yields $\Lambda = 0.32$ for an average beta function of $\beta_{\text{av}} = 5\,\text{m}$ and the FLASH parameters quoted in Chap. 5, so the 3D gain length is 32 % longer than the 1D gain length.

Quite a different parametrization has been proposed by Saldin et al. [12]. The power gain length[2] is written as

$$L_g = 1.19\left(\frac{I_A}{I_{\text{peak}}}\right)^{1/2} \frac{(\varepsilon_n\lambda_u)^{5/6}(1 + K^2/2)^{1/3}}{\lambda_\ell^{2/3}\widehat{K}}(1 + \delta) \tag{6.24}$$

with

$$\delta = \frac{262\,I_A}{I_{\text{peak}}}\,\frac{\varepsilon_n^{5/4}\sigma_\gamma^2}{\widehat{K}^2(1 + K^2/2)^{1/8}\lambda_\ell^{1/8}\lambda_u^{9/8}}\,. \tag{6.25}$$

Here $I_A = 4\pi\varepsilon_0 m_e c^3/e = 17.5\,\text{kA}$ is the so-called Alfvén current and I_{peak} is the peak current in the electron beam. The formula is only valid if the beta function is

[2] Note that in [12] the field gain length is computed which is a factor of two larger than the power gain length.

optimized according to

$$\beta_{\text{opt}} = 15.8 \left(\frac{I_A}{I_{\text{peak}}}\right)^{1/2} \frac{\varepsilon_n^{3/2} \lambda_u^{1/2}}{\widehat{K} \lambda_\ell} (1 + 8\delta)^{-1/3} . \tag{6.26}$$

The two parametrizations (6.22) and (6.24) are in reasonable agreement for LCLS and the European XFEL. From numerical FEL simulations, Kim and Xie [13] derived the following approximate formula for the saturation power

$$P_{\text{sat}} \approx 1.6 \, \rho_{\text{FEL}} \, P_{\text{beam}} \left(\frac{L_{g0}}{L_g}\right)^2 . \tag{6.27}$$

Here P_{beam} is the power of the electron beam.

6.5 FEL Radiation from Short Bunches

In an ultraviolet or X-ray FEL facility the electron bunches must be made very short because otherwise the required high peak current densities cannot be achieved. But also for scientific reasons a short bunch length is desirable since it permits the generation of femtosecond X ray pulses which are indispensable in the study of ultrafast processes.

In the 1D FEL theory of Chap. 4 we have totally disregarded the effects at the head and the tail of the bunch and basically treated the bunches as being infinitely long and having a periodic substructure. Here we consider what happens if the bunch length is finite and amounts to only a few hundred radiation wavelengths.

6.5.1 Velocities

In a high-gain FEL we have to distinguish different velocities which are all lower than the speed c of an electromagnetic wave in vacuum: the longitudinal speed $\bar{v}_z$ of the electrons, and the phase and the group velocity of the FEL light wave.

The terms phase and group velocity are familiar from other fields of physics. As an example we take a plane electromagnetic wave

$$E_x(z, t) = A \exp[i \, (kz - \omega t)] .$$

The phase and group velocities are

$$v_{\text{ph}} = \frac{\omega}{k} , \quad v_g = \frac{d\omega}{dk} .$$

For a light wave in free space the relationship between frequency and wave number is linear, $\omega = c\,k$, so phase and group velocity are equal, $v_{\rm ph} = v_g = c$. The situation is different in a dispersive medium like glass where the refractive index depends on frequency, $n = n(\omega)$. Then the wavelength is $\lambda' = \lambda/n$ and the wave number becomes

$$k' = n\,k = n(\omega)\,\omega/c\ .$$

The phase velocity

$$v_{\rm ph} = \frac{\omega}{k'} = \frac{c}{n} \tag{6.28}$$

is now different from the group velocity

$$v_g = \frac{d\omega}{dk'} = \left(\frac{dk'}{d\omega}\right)^{-1} = c\left(n + \omega\,\frac{dn}{d\omega}\right)^{-1} . \tag{6.29}$$

The group velocity is relevant for the propagation of wave packets, for example for digital data transmission through a glass fiber.

6.5.1.1 FEL Phase Velocity

The electron beam generates the FEL wave and has an influence on its propagation. We will demonstrate that the phase and group velocities are both less than the speed of light in vacuum and rather different from each other. The phase velocity of the FEL light wave is smaller than c because the phase $\varphi_E(z)$ of the complex field amplitude depends on z, as shown in Fig. 5.11. In the following we consider FEL seeding and assume for simplicity $\eta = 0$ and $k_p = 0$. The seed wave and the FEL wave are given by the expressions

$$E_{\rm seed}(z,t) = E_0\,\exp[i\,(kz - \omega t)]\,,$$
$$\tilde{E}_x(z,t) = \sum_{j=1}^{3} c_j\,\exp(\alpha_j z)\,\exp[i\,(kz - \omega t)] \quad \text{with} \quad c_j = E_0/3\,.$$

Using Eqs. (4.49) and (4.51) we write the three eigenvalues in the form

$$\alpha_1 = \frac{i + \sqrt{3}}{2\sqrt{3}\,L_{g0}}\,, \quad \alpha_2 = \frac{i - \sqrt{3}}{2\sqrt{3}\,L_{g0}}\,, \quad \alpha_3 = \frac{-2\,i}{2\sqrt{3}\,L_{g0}}\,. \tag{6.30}$$

For $0 \le z \ll L_{g0}$ we can make a first-order Taylor expansion of the exponential functions:

$$\exp(\alpha_j z) \approx 1 + \alpha_j z\,, \quad \sum_{j=1}^{3} \exp(\alpha_j z) \approx 3 + \underbrace{(\alpha_1 + \alpha_2 + \alpha_3)}_{0} = 3\,.$$

This means that the FEL wave is identical with the seed wave at the beginning of the undulator at $z = 0$, and consequently the FEL phase velocity is identical with the speed of light in vacuum:

$$v_{\mathrm{ph}}^0 = c \quad \text{for } 0 \le z \ll L_{g0}\,. \tag{6.31}$$

In the exponential gain regime get a different result. Here the field behaves as

$$\begin{aligned}\tilde{E}_x(z,t) &= (E_0/3)\,\exp[\alpha_1 z]\,\exp[i\,kz - i\omega t] \\ &= (E_0/3)\,\exp[\Re(\alpha_1)z]\,\exp[i\,(k + \Im(\alpha_1))z - i\omega t] \\ &\equiv (E_0/3)\,\exp[\Re(\alpha_1)z]\,\exp[i\,k'z - i\omega t]\,,\end{aligned} \tag{6.32}$$

where we have introduced a modified wave number k' by

$$k' = k + \Im(\alpha_1)\,. \tag{6.33}$$

The phase velocity is according to (6.28)

$$v_{\mathrm{ph}} = \frac{\omega}{k'} = \frac{\omega}{k + \Im(\alpha_1)} \approx c\left(1 - \frac{c\,\Im(\alpha_1)}{\omega}\right). \tag{6.34}$$

From Eq. (6.30) we find $\Im(\alpha_1) = 1/(2\sqrt{3}\,L_{g0})$. The phase velocity of the FEL wave in the exponential-growth regime becomes (writing now ω_ℓ instead of ω)

$$\boxed{v_{\mathrm{ph}} = c\left(1 - \frac{c}{2\sqrt{3}\,L_{g0}\,\omega_\ell}\right) = c\left(1 - \frac{\lambda_\ell}{4\pi\sqrt{3}\,L_{g0}}\right).} \tag{6.35}$$

The correction term is very small. For $\lambda_\ell = 4\,\mathrm{nm}$ and $L_{g0} = 0.44\,\mathrm{m}$ the phase velocity is $0.12\,\mathrm{m/s}$ lower than the speed of light in vacuum. Because of this tiny deviation from c, the FEL wave needs a long distance of $4\pi\sqrt{3}\,L_{g0} \approx 22\,L_{g0}$ to slip by just one wavelength with respect to a light wave in vacuum. This slow slippage is very important though as it is responsible for the FEL bucket motion shown in Fig. 5.13. The computed z-dependence of the phase velocity is plotted in Fig. 6.5.

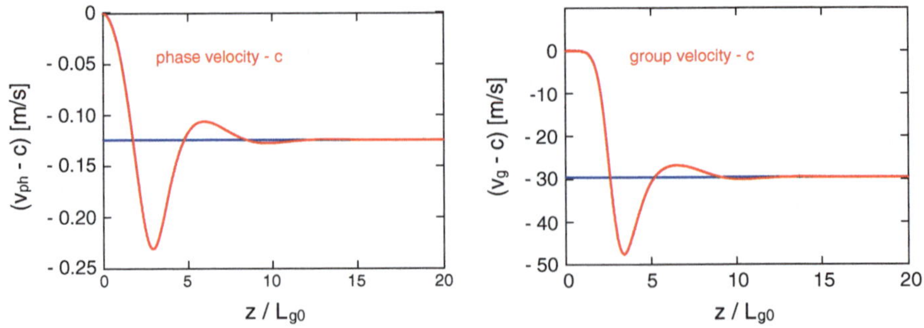

Fig. 6.5 *Left*: The difference between FEL phase velocity and speed of light, measured in m/s, is plotted as a function of the distance z traveled in the undulator. The asymptotic value (6.35) of the phase velocity is 0.12 m/s lower than c. *Right*: The difference between FEL group velocity and speed of light. The asymptotic value (6.37) of the group velocity is 30 m/s lower than c. Parameters used in the computation: $\lambda_\ell = 4\,\text{nm}$ and $L_{g0} = 0.44\,\text{m}$.

6.5.1.2 FEL Group Velocity

Of particular interest for a SASE FEL is the velocity of a "spike" in the FEL wave that extends over many optical wavelengths. The spike is represented by a wave packet propagating with the group velocity. The group velocity is equal to c for $0 \le z \ll L_{g0}$, but with increasing z it drops quickly to an asymptotic value, as shown in Fig. 6.5. In the exponential-growth regime the group velocity is according to (6.29)

$$v_g = \frac{d\omega}{dk'} = \left(\frac{dk'}{d\omega}\right)^{-1} = \left(\frac{dk}{d\omega} + \frac{d\Im(\alpha_1)}{d\omega}\right)^{-1}$$
$$= \left(\frac{1}{c} + \frac{d\Im(\alpha_1)}{d\omega}\right)^{-1} \approx c\left(1 - c\,\frac{d\Im(\alpha_1)}{d\omega}\right). \qquad (6.36)$$

The derivative $d\Im(\alpha_1)/d\omega$ is computed in Sect. 10.4, see Eq. (10.96):

$$\frac{d\Im(\alpha_1)}{d\omega} \approx \frac{2k_u}{3\omega_r}\,.$$

From (6.36) and $c\,k_u = \omega_r(1 + K^2/2)/(2\gamma_r^2)$ follows that the group velocity in the exponential-growth regime is given by the expression

$$\boxed{v_g = c\left(1 - \frac{1}{3\gamma_r^2}\left[1 + \frac{K^2}{2}\right]\right).} \qquad (6.37)$$

Remarkably, the group velocity of a spike in the electron beam current density is identical with the optical group velocity. The proof is as follows. The current density is

$$\tilde{j}_z(\psi, z) = j_0 + \tilde{j}_1(z)\,\exp(i\,\psi) \quad \text{with} \quad \psi = (k_\ell + k_u)z - \omega_\ell t\,.$$

According to Eq. (4.21) the complex amplitude $\tilde{j}_1$ is proportional to the derivative of $\tilde{E}_x$. In the exponential gain regime, the modulated current density $\tilde{j}_1$ has therefore the same z dependence as $\tilde{E}_x$, namely $\exp(\alpha_1 z)$:

$$\tilde{j}_1\,\exp(i\,\psi) \propto \exp[\Re(\alpha_1)z]\,\exp[i\,\Im(\alpha_1)z]\exp[i(k_\ell + k_u)z - i\,\omega_\ell t]\,.$$

The phase of the modulated current density differs from that of the FEL field only by the term $k_u z$. Now $k_u = 2\pi/\lambda_u$ is a constant, independent of frequency. Application of formula (6.36) yields thus the exactly same group velocity as in the FEL wave. This equality has a deep physical relevance: a bump in the electron charge distribution is associated with a corresponding bump in the FEL field due to the FEL gain process, and hence one would expect that both bumps move with the same speed, which they indeed do.

The group velocity is considerably lower than the phase velocity. We have seen above that in an undulator of the length $L_u = 22\,L_{g0}$ the phase of a harmonic FEL wave slips by one optical wavelength with respect to a plane electromagnetic wave. In contrast to this an FEL wave packet slips by a few hundred optical wavelengths. On the other hand, the group velocity is larger than the average longitudinal speed of the electrons which is according to Eq. (2.11)

$$\bar{v}_z = c\left(1 - \frac{1}{2\gamma_r^2}\left[1 + \frac{K^2}{2}\right]\right).$$

The difference is

$$\boxed{v_g - \bar{v}_z = \frac{c}{6\gamma_r^2}\left[1 + \frac{K^2}{2}\right].} \tag{6.38}$$

From this we conclude that spikes (wave packets) in the FEL wave, or spikes in the electron charge distribution, will slip forward inside the electron bunch during the motion through the undulator.

6.5.2 *Slippage Effects in Short Bunches*

We consider an example which illustrates very nicely the slippage effects in free-electron lasers that are driven by short electron bunches. The bunch considered has a narrow leading spike and a long tail. The local current I_0 as a function of the internal bunch coordinate ζ is depicted in Fig. 6.6. The peak current is $I_{\text{peak}} = 1600$ A. We assume that the lasing process is started by seed radiation with an amplitude of $E_0 = 5$ MV/m. The simulation of the FEL process is done using the coupled

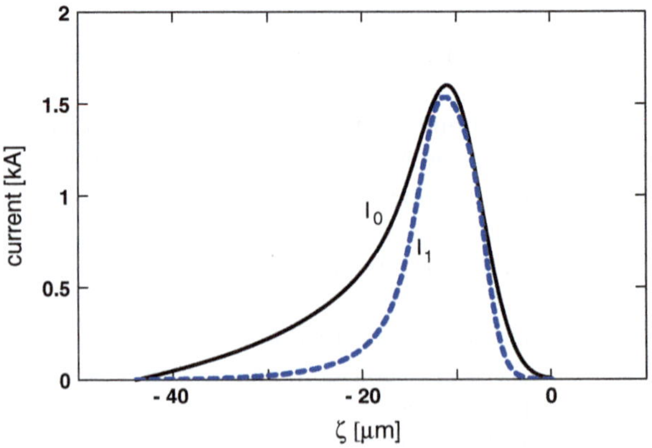

Fig. 6.6 An electron bunch with a narrow leading spike and a long tail. The local "direct current" I_0 is plotted as a function of the internal bunch coordinate ζ (*solid black curve*). The bunch head is on the right side. The modulation current I_1 is shown by the *dashed blue curve*. The longitudinal position in the undulator magnet is $z = 18\,L_{g0}$.

first-order Eqs. (10.63a,10.63b,10.63c,10.63d) in their generalized form, including the dependencies on the internal bunch coordinate ζ. The typical FLASH parameters quoted in Chap. 5 are used.

At a position of $z = 18\,L_{g0}$ in the undulator the maximum of the modulation current has almost reached the peak current, but the width of the curve $I_1(t)$ is narrower than the bunch itself whose profile is described by the curve $I_0(t)$. This is due to the fact that the FEL gain depends on the beam current and is suppressed in the tail of the bunch.

In Fig. 6.7 the electric field of the light wave (in units of 10^{10}V/m) and the normalized modulation current $I_1/I_{\rm peak}$ are shown at various positions in the undulator. Close to the end of the exponential gain regime at $z = 18\,L_{g0}$ the modulation current I_1 has reached the level of the peak current and the electric field of the light wave has grown to 8000 MV/m. Here the longitudinal profile of the field resembles closely the shape of the I_1 distribution inside the bunch, but is a bit narrower than the bunch itself.

Next we look into the saturation regime. At $z \geq 20\,L_{g0}$, the modulation current develops a structure with two or more peaks that is quite different from the bunch shape. The electric field amplitude of the light wave is also characterized by a complicated structure. Narrow wave packets of the FEL field escape from the bunch and move away from the bunch head. This is clear evidence for the slippage effect. When an FEL wave packet has slipped away from the bunch it will move with the speed c of light in vacuum, and its magnitude will remain invariant because the overlap with the electron beam is no longer existent and the FEL gain process has come to an end.

It is important to note that the spikes are by no means identical with the microbunches. Rather they are wave packets extending over many optical wavelengths. Inside the bunch they propagate with the FEL group velocity (6.37) and are

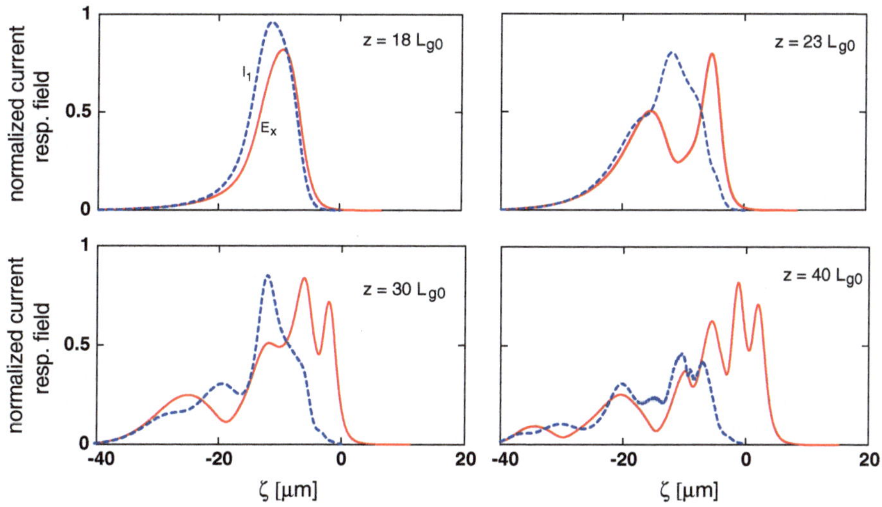

Fig. 6.7 FEL radiation from the bunch shown in Fig. 6.6. The electric field of the FEL wave is plotted as a function of the internal bunch coordinate ζ (*continuous red curves*). The unit is 10^{10} V/m $=$ 10 GV/m. The normalized modulation current I_1/I_{peak} is shown by the *dashed blue curves*. The evolution of modulation current and FEL field is displayed at various longitudinal positions in the undulator magnet: $z = 18\,L_{g0}$, $23\,L_{g0}$, $30\,L_{g0}$ and $40\,L_{g0}$.

thus faster than the microbunches which move with a speed that is very close to the average longitudinal speed $\bar{v}_z$ of the electrons.

6.5.2.1 Longitudinal Structure of the FEL Pulses

The above figures show that the simulated FEL pulses contain amazingly short wave packets. What is the physical origin of these sharp substructures? We want to demonstrate that this is nothing else but the oscillatory behavior of the FEL power in the saturation regime that we observed earlier in Chap. 5. In order to see this we go back to Fig. 5.4 where the computed FEL power on a logarithmic scale has been plotted against z/L_{g0}. The same data are used to show in Fig. 6.8a the z dependence of the absolute magnitude of the electric light wave field $\left|\tilde{E}_x(z)\right|$ on a linear scale. Here the oscillations reflect obviously the back-and-forth energy exchange between electron bunch and light wave. Converting to the internal bunch coordinate ζ and taking into consideration that the bunch head enters the saturation regime earlier than the tail (which implies a sign reversal) one obtains similar sharp time spikes as seen in Fig. 6.7.

The question arises whether these sharp peaks are to be expected in a realistic FEL or whether they are just an artifact of our oversimplified one-dimensional FEL model. The second alternative turns out to be correct. According to computations with three-dimensional FEL codes, the gain oscillations in the saturation regime are much

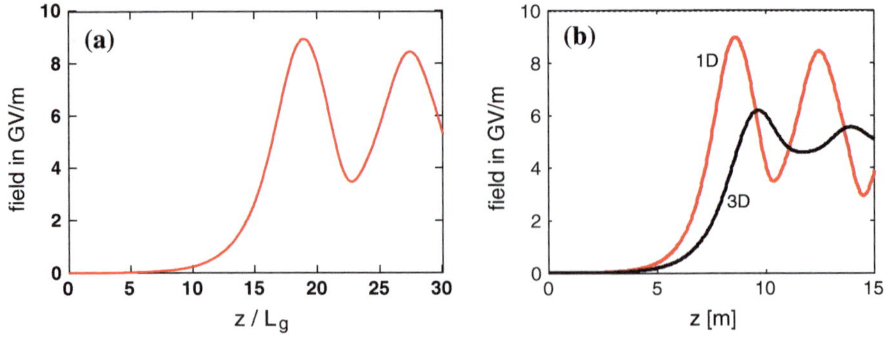

Fig. 6.8 **a** The magnitude of the electric light wave field on a linear scale, plotted as a function of the longitudinal position z in the undulator. The parameters are the same as in Fig. 5.4. **b** Comparison of the field rise in the 1D and the 3D FEL theory.

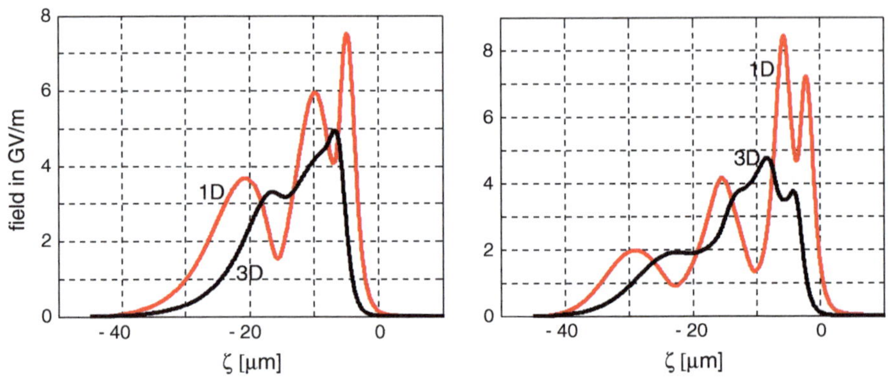

Fig. 6.9 Comparison of the FEL pulse shape in the 1D theory (*red curve*) and the full 3D theory (*black curve*) at two positions in the undulator. We thank Igor Zagorodnov of DESY for carrying out the 3D calculations using the code GENESIS.

less pronounced than predicted by the 1D theory, see Fig. 6.8b. This is in fact easy to understand. In the 1D model the density of the electron beam and the amplitude of the light wave do not depend on the radial coordinate $r = \sqrt{x^2 + y^2}$ while in reality both charge density and field amplitude drop with increasing distance from the beam axis. The reduced coupling strength at $r > 0$ implies that the saturation regime is reached later for off-axis electrons than for on-axis electrons. Consequently, we can expect that the gain oscillations will be smeared out for realistic electron and photon beams. This smearing will also wash out the spike structure of the FEL pulses as demonstrated in Fig. 6.9. In addition, optical diffraction plays a larger role in the saturation regime and reduces the electric field close to the axis.

6.6 Superradiance

Bonifacio and others [14] predicted the possibility of a reduction of the light pulse length to values below the electron bunch length if the exponential gain regime has been passed. The term *superradiance* is used to describe this pulse shortening regime in which the radiated FEL power is proportional to the square of the number of electrons. Superradiance was observed [15] in a single-pass high-gain FEL which was seeded at 800 nm by a titanium-sapphire laser. Using frequency resolved optical gating the FEL pulse duration was measured to be 82 fs (rms) while the seed pulse was 150 fs long, and the electron bunch was even more than 1000 fs long.

The interpretation of superradiance is based on the assumption that the radiation pulse slips forward inside the bunch, owing to the fact that the optical group velocity is larger than the longitudinal speed of the electrons, and that this light pulse therefore withdraws energy from "fresh" electrons that have not yet been microbunched. In the superradiant regime the FEL power is expected to grow quadratically with the distance z traveled in the undulator.

References

1. W.M. Fawley, Report No. LBNL-49625, Lawrence Berkeley Laboratory (2002)
2. S. Reiche, GENESIS 1.3, a fully 3D time-dependent FEL simulation code. Nucl. Instrum. Meth. A **429**, 243 (1999)
3. E.L. Saldin, E.A. Schneidmiller, M.V. Yurkov, FAST: a three-dimensional time-dependent FEL simulation code. Nucl. Instrum. Meth. A **429**, 233 (1999)
4. E.L. Saldin, E.A. Schneidmiller, M.V. Yurkov, *The Physics of Free Electron Lasers* (Springer, Berlin, 2000)
5. Z. Huang, K.-J. Kim, Review of x-ray free-electron laser theory. Phys. Rev. ST Accel. Beams. **10**, 034801 (2007)
6. D.A. Edwards, M.J. Syphers, *An Introduction to the Physics of High Energy Accelerators* (John Wiley, New York, 1993)
7. E.T. Scharlemann, Wiggle plane focusing in linear wigglers. J. Appl. Phys. **58**, 2154 (1985)
8. E.T. Scharlemann, A.M. Sessler, J.S. Wurtele, Optical guiding in a free-electron laser. Phys. Rev. Lett. **54**, 1925 (1985)
9. H.P. Freund, T.M. Antonsen, *Principles of Free-Electron Lasers*, 2nd edn. (Chapman and Hall, London, 1996)
10. Z. Huang, K.-J. Kim, Transverse and temporal characteristics of a high-gain free-electron laser in the saturation regime. Nucl. Instrum. Meth. A. **483**, 504 (2002)
11. M. Xie, Exact and variational solutions of 3D eigenmodes in high gain FELs. Nucl. Instrum. Meth. A **445**, 59 (2000)
12. E.L. Saldin, E.A.Schneidmiller, M.V. Yurkov, Design Formulas for VUV and X-Ray FELs, in *Proceedings of FEL Conference*, 2004
13. K.-J. Kim, M. Xie, Self-amplified spontaneous emission for short wavelength coherent radiation. Nucl. Instrum. Meth. A **331**, 359 (1993)
14. R. Bonifacio et al., Spectrum, temporal structure, and fluctuations in a high-gain free-electron laser starting from noise. Phys. Rev. Lett. **73**, 70 (1994)
15. T. Watanabe et al., Experimental characterization of superradiance in a single-pass high-gain laser-seeded free-electron laser. Phys. Rev. Lett. **98**, 038402 (2007)

Chapter 7
Self-Amplified Spontaneous Emission and FEL Seeding

For wavelengths in the extreme ultraviolet and X-ray regime the start-up of the FEL process by seed radiation is hampered by the lack of lasers with the desired wavelength. Seeding by a high harmonic of an optical or infrared laser is a possibility which has been realized in recent years, see Sect. 7.4, where also other seeding schemes are described. The process of Self-Amplified Spontaneous Emission (SASE) permits the start-up of lasing at an arbitrary wavelength, without the need of external radiation. The SASE mechanism was proposed and theoretically explored in the early 1980s [1–4] but it took 20 years before it could be verified experimentally at visible and ultraviolet wavelengths.

The most intuitive explanation of SASE is that the electrons produce spontaneous undulator radiation in the first section of a long undulator magnet which serves then as seed radiation in the main part of the undulator. An alternative explanation follows from the observation that the FEL can also be started by a periodic charge density modulation in the electron beam, as discussed in Sect. 5.3. The bunches coming from the accelerator do not possess such a modulation at the light wavelength. But due to the fact that they are composed of a large number of randomly distributed electrons a white noise spectrum is generated which has a spectral component within the FEL bandwidth (see Sect. 10.5.1). This component will be amplified as shown in Sect. 5.3.

The above two interpretations of SASE are physically equivalent: seeding by spontaneous undulator radiation or FEL start-up by the proper Fourier component of the stochastic density modulation in the electron beam. Randomness is obviously essential in the second model of the SASE process, but it is equally important in the first model. It must be noted that the emission of undulator radiation by a bunch much longer than the light wavelength would be impossible if the longitudinal particle distribution were perfectly uniform, in the extreme case, if the electron beam current would be a direct current. A perfect dc current moving on a sinusoidal orbit through the undulator magnet cannot emit any radiation because there are no oscillating charges. Likewise, a perfect dc current circulating in an electron synchrotron or storage ring would be unable to emit ordinary synchrotron radiation.

P. Schmüser et al., *Free-Electron Lasers in the Ultraviolet and X-Ray Regime*, Springer Tracts in Modern Physics 258, DOI: 10.1007/978-3-319-04081-3_7,

7.1 Computation of the SASE Process in the 1D Theory

7.1.1 Solution of the Third-Order Equation

In this section the equivalent current density modulation $\tilde{j}_1$ arising from the random time distribution of the electrons in the bunch is used as an input for calculating the time evolution of the FEL power by means of the method discussed in Sect. 5.3. The initial conditions are given by Eq. (5.15)

$$\begin{pmatrix} \tilde{E}_x(0) \\ \tilde{E}_x'(0) \\ \tilde{E}_x''(0) \end{pmatrix} = \begin{pmatrix} 0 \\ -1 \\ i\,2k_u\eta \end{pmatrix} \frac{\mu_0 c\,\widehat{K}}{4\gamma_r}\,\tilde{j}_1(0)\,.$$

It is convenient to factor out the driving term $\mu_0 c\,\widehat{K}\,\tilde{j}_1(0)/(4\gamma_r)$ by introducing new coefficients d_j

$$\begin{pmatrix} d_1 \\ d_2 \\ d_3 \end{pmatrix} = \begin{pmatrix} 1 & 1 & 1 \\ \alpha_1 & \alpha_2 & \alpha_3 \\ \alpha_1^2 & \alpha_2^2 & \alpha_3^2 \end{pmatrix}^{-1} \cdot \begin{pmatrix} 0 \\ -1 \\ i\,2k_u\eta \end{pmatrix}. \tag{7.1}$$

Then the field amplitude becomes

$$\tilde{E}_x(z) = \sum_{j=1}^{3} d_j \exp(\alpha_j z) \cdot \frac{\mu_0 c\,\widehat{K}}{4\gamma_r}\,\tilde{j}_1(0)\,. \tag{7.2}$$

According to Eq. (10.116) in Sect. 10.5.1 the equivalent modulated current density resulting from shot noise is given by the formula

$$\tilde{j}_1(0) = \sqrt{\frac{e\,I_0\,\Delta\omega}{\pi}}\,\frac{1}{A_b}\,, \tag{7.3}$$

where I_0 is the absolute magnitude of dc electron beam current, A_b is the beam cross section, and $\Delta\omega$ is the FEL bandwidth.

First an approximate computation is made with a constant bandwidth. To achieve saturation one needs about 20 gain lengths, see Fig. 7.1 below. Using Eq. (5.14) we compute a bandwidth of $\Delta\omega \approx 2.4\,\omega_\ell\,\rho_{\mathrm{FEL}}$ and substitute this in Eq. (7.3). The central frequency is taken as $\omega = \omega_r$, hence $\eta = 0$ in Eq. (7.1). The FEL power as a function of the position z in the undulator is computed with Eq. (7.2)

$$P_{\mathrm{cb}}(z) = \frac{A_b}{\mu_0 c}\left|\tilde{E}_x(z)\right|^2 = \frac{\mu_0 c\,\widehat{K}^2}{16\gamma_r^2}\left|\sum_j d_j \exp(\alpha_j z)\right|^2 \frac{e\,I_0}{\pi A_b}\,\underbrace{2.4\,\rho_{\mathrm{FEL}}\,\omega_\ell}_{\Delta\omega}\,. \tag{7.4}$$

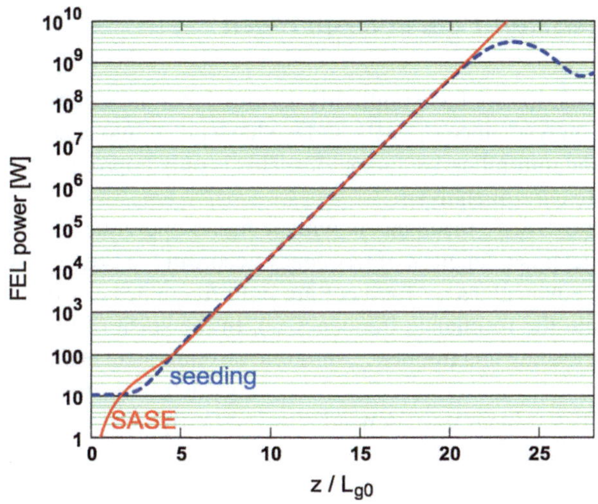

Fig. 7.1 Computed SASE FEL power as a function of z/L_{g0} (*continuous red curve*). The start-up is provided by the electron current density modulation with the period λ_ℓ which is due to the random distribution of the electrons in the bunch. For comparison the power rise of a seeded FEL is also shown, choosing a seeding field of $E_0 = 0.5$ MV/m (*dashed blue curve*). Here the gain curve has been computed by means of the coupled first-order equations so that the FEL saturation is included.

Here we have assumed that electron beam and light beam have the same cross section A_b. The subscript "cb" stands for constant bandwidth. The FEL power as a function of z/L_{g0} is shown in Fig. 7.1. In this model calculation we have used typical parameters of the free-electron laser FLASH (see Chap. 8).

It is interesting to compare the power rise curve of a SASE FEL to that of a seeded FEL. The field amplitude E_0 of the seed wave has been adjusted to yield the same power level in the exponential regime as obtained in the SASE FEL. For FLASH at an electron energy of 500 MeV we find $E_0 = 0.5$ MV/m. This computation permits us to determine the minimum seed laser field which is needed to be well above the SASE level. In the present example, seeding will dominate for an initial field value E_0 exceeding a few MV/m. We come back to this important issue in Sect. 7.1.3.

We learn from Fig. 7.1 that saturation is reached after about 20 gain lengths, so the *saturation length* is

$$L_{\text{sat}} \approx 20 L_{g0} .$$

This is in fact a rather typical number. In Ref. [5] the following formula is presented for the saturation length of a SASE X-ray FEL

$$L_{\text{sat}} \approx \frac{\lambda_u}{\rho_{\text{FEL}}} = 4\pi\sqrt{3} L_{g0} = 21.8\, L_{g0} . \tag{7.5}$$

7.1.2 Transmission Function Approach

In the previous section we have implicitly assumed that the eigenvalues α_j are given by their values on resonance ($\eta = 0$), and we have taken into account the bandwidth of the FEL by the multiplicative factor $\Delta\omega$. This is only approximately correct. Now we make a more careful analysis.

We study the SASE mechanism with a mono-energetic electron beam. The detuning parameter is a function of frequency according to Eq. (5.8). Our first goal is to compute the eigenvalues α_j for a certain light frequency ω which will in general be different from the resonance frequency ω_r. In the spirit of the argumentation in Sect. 5.2.1 we translate this frequency shift into an equivalent energy shift of the electron beam. The eigenvalues for a beam with energy detuning are given by Eq. (10.93) in Sect. 10.4. These eigenvalues depend now implicitly on frequency.

We define a *transmission function* relating current and field in frequency domain

$$H(\omega, z) = \left(e^{\alpha_1 z}\ e^{\alpha_2 z}\ e^{\alpha_3 z}\right) \cdot \begin{pmatrix} 1 & 1 & 1 \\ \alpha_1 & \alpha_2 & \alpha_3 \\ \alpha_1^2 & \alpha_2^2 & \alpha_3^2 \end{pmatrix}^{-1} \cdot \begin{pmatrix} 0 \\ -1 \\ i\,2k_u\eta \end{pmatrix}. \tag{7.6}$$

The spectral component of the electric field at the position z is

$$\tilde{E}_x(\omega, z) = \frac{\mu_0 c\,\widehat{K}}{4\gamma_r} H(\omega, z) J(\omega)\,, \quad J(\omega) = \sqrt{\frac{e\,I_0}{\pi}} \frac{1}{A_b}\,, \tag{7.7}$$

where $J(\omega)$ is the spectral current density resulting from shot noise, see Eq. (10.115) in Sect. 10.5.1. The radiated power is

$$P(z) = \frac{A_b}{\mu_0 c} \int_0^\infty \left|\tilde{E}_x(\omega, z)\right|^2 d\omega = \frac{\mu_0 c\,\widehat{K}^2 e\,I_0}{16\pi\gamma_r^2 A_b} \int_0^\infty |H(\omega, z)|^2\,d\omega\,. \tag{7.8}$$

The power computed from this equation is compared in Fig. 7.2 with the power predicted by the constant-bandwidth formula. Good agreement is observed at large values of z/L_{g0} but the transmission function automatically incorporates the variable bandwidth and yields thus higher FEL power in the first section of the undulator.

In the exponential-growth regime, the transmission function increases exponentially along the undulator axis, and its frequency dependence can be approximated by a Gaussian

$$H(\omega, z) \propto \exp\left(\frac{z}{2\,L_{g0}}\right) \exp\left(-\frac{(\omega-\omega_r)^2}{4\sigma_\omega^2}\right). \tag{7.9}$$

This follows from the approximate form (5.13) of the FEL gain function. The z-dependent rms frequency bandwidth is given by Eq. (5.14)

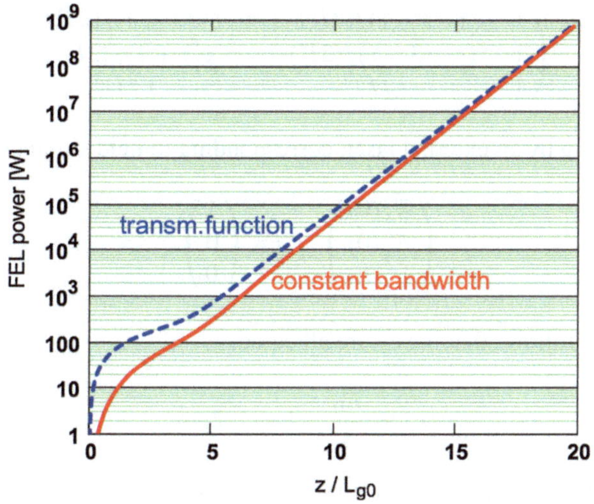

Fig. 7.2 *Dashed blue curve*: Computed SASE FEL power as a function of z/L_{g0} using the transmission-function method, Eq. (7.8). *Solid red curve*: power $P_{cb}(z)$ computed with the constant-bandwidth formula (7.4).

$$\sigma_\omega = \sigma_\omega(z) = 3\sqrt{2}\,\rho_{\rm FEL}\;\omega_\ell\sqrt{\frac{L_{g0}}{z}}\,.$$

7.1.3 Comparison of FEL Start-Up by Seeding or by SASE

In order to compare the two different start-up mechanisms, seeding or SASE, we determine the coefficients c_j in the eigenfunction expansion

$$\tilde{E}_x(z) = \sum_j c_j \exp(\alpha_j z)$$

by the matrix Equation (4.54)

$$\begin{pmatrix} c_1 \\ c_2 \\ c_3 \end{pmatrix} = \boldsymbol{\mathcal{A}}^{-1} \cdot \begin{pmatrix} \tilde{E}_x(0) \\ \tilde{E}_x'(0) \\ \tilde{E}_x''(0) \end{pmatrix}.$$

We take the simplest form of the matrix $\boldsymbol{\mathcal{A}}^{-1}$, see Eq. (4.56):

$$\mathcal{A}^{-1} = \frac{1}{3} \cdot \begin{pmatrix} 1 & (\sqrt{3}-i)/(2\Gamma) & (-i\sqrt{3}+1)/(2\Gamma^2) \\ 1 & (-\sqrt{3}-i)/(2\Gamma) & (i\sqrt{3}+1)/(2\Gamma^2) \\ 1 & i/\Gamma & -1/\Gamma^2 \end{pmatrix} . \tag{7.10}$$

In case of seeding by an external laser, corresponding to field stimulation, the initial condition is

$$\begin{pmatrix} \tilde{E}_x(0) \\ \tilde{E}'_x(0) \\ \tilde{E}''_x(0) \end{pmatrix} = E_0 \begin{pmatrix} 1 \\ 0 \\ 0 \end{pmatrix} ,$$

and from (7.10) follows that the c_j are all equal,

$$c_j = \frac{E_0}{3}, \quad j = 1, 2, 3.$$

In case of SASE we have current stimulation caused by shot noise. Then the initial condition reads for $\eta = 0$

$$\begin{pmatrix} \tilde{E}_x(0) \\ \tilde{E}'_x(0) \\ \tilde{E}''_x(0) \end{pmatrix} = -\frac{\mu_0 c \widehat{K}}{4\gamma_r} \tilde{j}_1 \begin{pmatrix} 0 \\ 1 \\ 0 \end{pmatrix} .$$

Applying (7.10) we find

$$|c_j| = \frac{1}{3\Gamma} \frac{\mu_0 c \widehat{K}}{4\gamma_r} \tilde{j}_1 .$$

The equivalent input field is therefore

$$E_{\text{equiv}} = \frac{\mu_0 c \widehat{K}}{4\gamma_r \Gamma} \tilde{j}_1 = \frac{\mu_0 c \widehat{K}}{4\gamma_r \Gamma} \frac{\sqrt{e I_0 \Delta\omega}}{\sqrt{\pi} A_b} . \tag{7.11}$$

With the FLASH parameters used in Sect. 7.1.1 we obtain an equivalent input field of 0.54 MV/m, in good agreement with the value of 0.5 MV/m used in Fig. 7.1. Only in the case that the seeding field E_0 exceeds the equivalent field E_{equiv} by a sufficient margin one can expect that the properties of the FEL output radiation such as pulse length, coherence and bandwidth are determined by the seed radiation rather than by the SASE process, which is always present and competes with seeding. The required seeding field level has to be maintained over at least one gain length. This is an important constraint on the beam quality and power of the seed radiation which becomes quite demanding if one wants to use a very high harmonic of optical or infrared laser radiation.

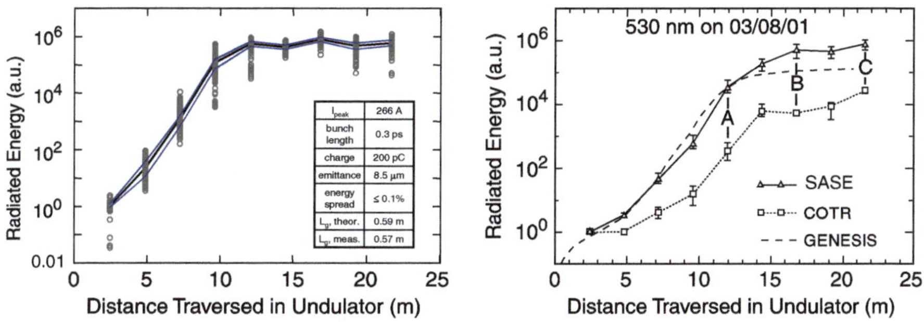

Fig. 7.3 SASE radiation at 530 nm in the low-energy undulator test line LEUTL. *Left*: The FEL intensity is shown as a function of the distance traversed in the undulator (courtesy of S. Milton). *Right*: The energy of SASE FEL radiation and of coherent optical transition radiation as a function of distance traversed in the undulator [9]. The *dashed curve* shows the prediction of the FEL code GENESIS. Figure adapted with permission from [9]. ©2002 by the American Physical Society.

7.2 Experimental Results on SASE FEL Radiation

7.2.1 Detection of SASE, Exponential Gain and Saturation

The first experimental verification of the SASE mechanism was achieved in 1998 in the infrared wavelength range [6, 7]. SASE FEL operation in the visible and near-ultraviolet range was accomplished in 2001 at the low-energy undulator test line LEUTL at Argonne National Laboratory near Chicago, USA [8]. The gain curve measured at 530 nm is shown in Fig. 7.3, similar data were obtained at 385 nm. The exponential rise and saturation of the FEL intensity was observed both at the visible and the ultraviolet wavelength.

The LEUTL team made a very nice experimental verification of the microbunching associated with the high-gain FEL process by detecting coherent optical transition radiation[1] (COTR) at the FEL wavelength. The data in Fig. 7.3 demonstrate convincingly the evolution of microbunches with the periodicity given by the light wavelength: the COTR power exhibits a similar exponential rise as the FEL power.

In 2001 a successful SASE experiment was carried out at DESY in Hamburg at the vacuum-ultraviolet (VUV) wavelength of 109 nm [10]. In the next series of measurements, wavelengths between between 80 and 180 nm were covered and FEL saturation was established [11, 12]. The data at 98 nm have already been presented in Fig. 4.2. After an energy upgrade to 1250 MeV and an increase of the undulator

[1] Transition radiation is produced when relativistic particles cross the boundary between two media of different refractive indices. The radiation emitted in backward direction is in the visible and infrared range. Optical transition radiation (OTR) is frequently applied at electron accelerators to obtain images of the beam cross section, see Sect. 8.9. The radiation is usually incoherent because in most cases the bunch is much longer than the optical wavelength. Coherent optical transition radiation with a much increased intensity will be generated if the bunch possesses a periodic density modulation with the period being equal to the optical wavelength.

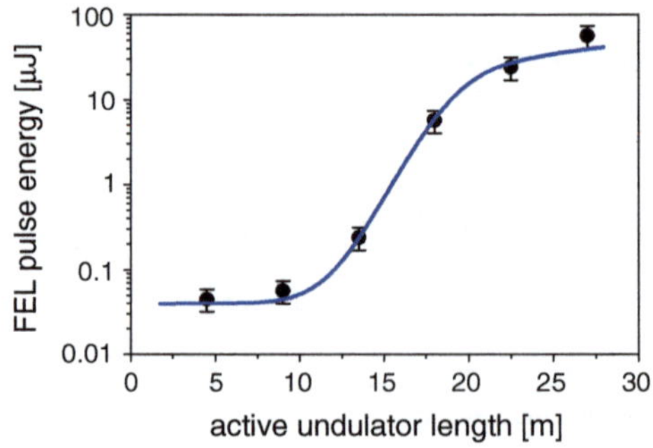

Fig. 7.4 Experimental evidence for exponential gain and FEL saturation in the extreme ultraviolet at $\lambda_\ell = 13.7\,\text{nm}$ [13]. The average light pulse energy is plotted as a function of the active undulator length. Note that this light pulse energy is in fact the sum of the FEL pulse energy and the contribution from spontaneous undulator radiation entering the light detector. The active undulator length z is varied in the experiment by switching on kicker magnets at selected positions inside the long undulator system, thereby destroying the good overlap between electron and photon beam in the downstream section of the undulator and inhibiting further FEL gain. The intensity of the spontaneous undulator radiation is hardly affected by the induced small orbit distortions.

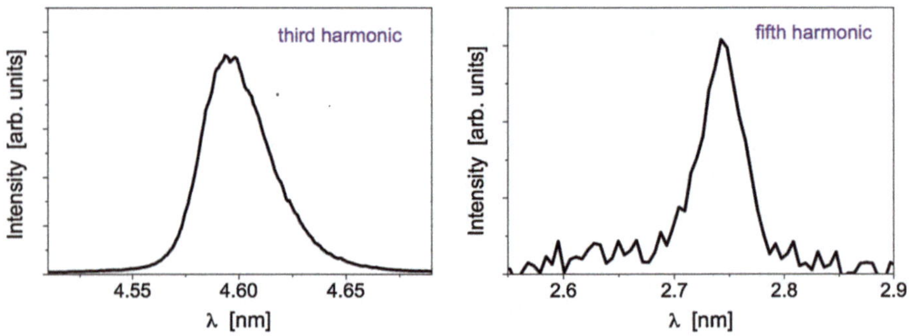

Fig. 7.5 Spectra of the third harmonic (*left*) and fifth harmonic (*right*) of 13.7 nm FEL radiation [13]. For an average pulse energy of 40 µJ at the first harmonic, a pulse energy of 0.25 µJ was measured at the third harmonic (4.6 nm) and of 10 nJ at the fifth harmonic (2.75 nm).

length from 13.5 to 27 m the FEL was renamed into FLASH. This FEL user facility will be described in Chap. 8.

The data at a fundamental wavelength of 13.7 nm [13] are presented in Fig. 7.4. Analysis of the exponential part of the gain curve yields a power gain length of $L_g = 1.25 \pm 0.15\,\text{m}$. Quite remarkable is the observation of the third and fifth harmonic at 4.6 and 2.75 nm, respectively, see Fig. 7.5. The fifth harmonic is within the so-called *water window*, a spectral region in the extreme ultraviolet where water becomes transparent. Radiation in this region is crucially important for the investigation of biological matter. Higher harmonics of the 32 nm line have also been seen [14]. In this measurement the second harmonic at 15.9 nm was observed, but with a twenty

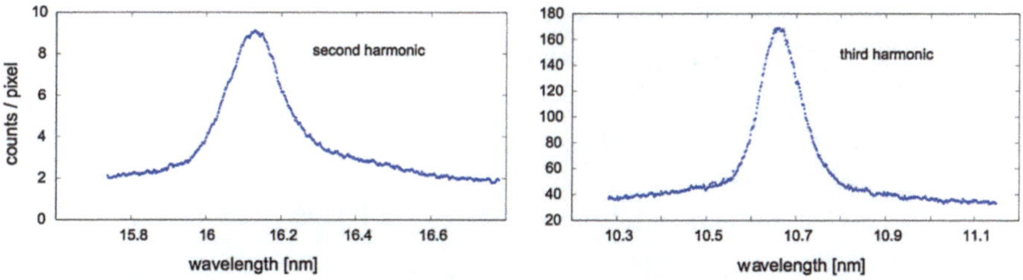

Fig. 7.6 Spectra of the second harmonic *(left)* and third harmonic *(right)* of 32 nm FEL radiation [14].

times lower intensity than the third harmonic at 10.5 nm (Fig. 7.6). Remembering the discussion in Sect. 3.5 the suppression of the even harmonics is no surprise.

SASE radiation at 840 nm and at the second and third higher harmonic have been reported from the VISA FEL at Brookhaven National Laboratory (BNL), USA [15]. Both SASE and seeded FEL lasing have been accomplished at wavelengths of 266 and 400 nm in the Deep Ultra-Violet (DUV) FEL at the National Synchrotron Light Source at BNL [16]. SASE radiation at 49 nm was observed in 2006 at the test accelerator SCSS of the Japanese X-ray FEL project [17]. Recent results on SASE radiation in the X-ray regime will be presented in Chap. 9.

7.2.2 Transverse Coherence of SASE Radiation

As shown in Sect. 10.3, the fundamental Gaussian mode TEM_{00} has its highest intensity on the beam axis while the higher TEM_{mn} modes extend to larger radial distances and some of them even vanish on the axis. With increasing length in the undulator, the fundamental TEM_{00} mode will therefore grow faster than the other modes, owing to its superior overlap with the electron beam. This process is called mode competition. When the saturation regime is approached the fundamental mode will usually dominate and the FEL radiation will possess a high degree of transverse coherence. This has indeed been verified by double-slit diffraction experiments. The data shown in Fig. 7.7 prove that almost full transverse coherence has been achieved at FEL wavelengths of 100 and 7 nm. This feature is of extreme importance for a large class of experiments.

It is interesting to note [5] that a SASE FEL can reach almost full transverse coherence even if the emittance criterion $\varepsilon \leq \lambda_\ell/(4\pi)$ is not fulfilled. Numerical calculations for the Linac Coherent Light Source LCLS, made by S. Reiche with his code GENESIS, reveal that in the early part of the undulator several transverse FEL beam modes will be excited if the electron beam cross section is too large, see the left picture in Fig. 7.8. The fundamental TEM_{00} mode grows faster than the

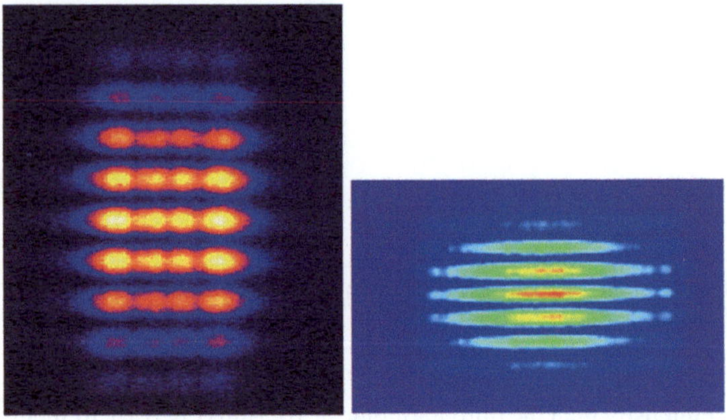

Fig. 7.7 Measured double-slit diffraction patterns. *Left*: at a fundamental wavelength of $\lambda_1 = 100\,\text{nm}$ [18, 19]. The slit separation is 0.5 mm. *Right*: at $\lambda_1 = 7\,\text{nm}$ (courtesy of Marion Kuhlmann, DESY).

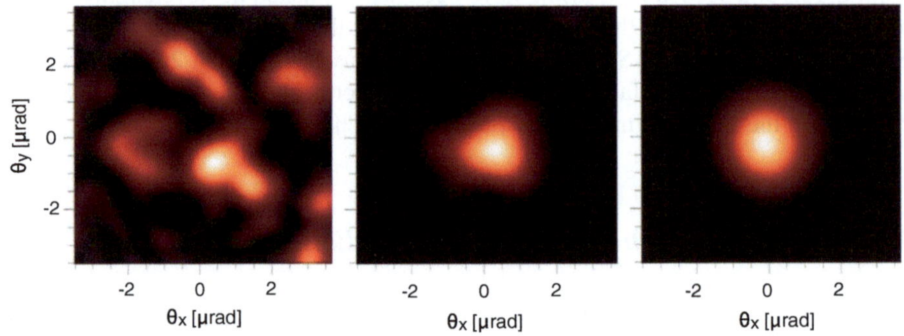

Fig. 7.8 Computed evolution of the transverse angular distribution of the FEL light in LCLS at different longitudinal positions z in the undulator. *Left*: $z = 4\,L_g$, *center* $z = 10\,L_g$, *right*: $z = 19\,L_g$. (Courtesy of S. Reiche).

other modes and exceeds their intensity already at 10 gain lengths. At the end of the exponential gain regime ($z = 19\,L_g$), the TEM_{00} mode dominates and the FEL beam has acquired almost full transverse coherence. This is a nice illustration of "mode cleaning".

Deep in the saturation regime of the fundamental mode, the higher modes are not yet saturated and may continue to grow with increasing z, depending on the energy spread acquired by the electron beam upon passing the undulator. Consequently, the degree of transverse coherence may drop below 90 %.

7.3 Statistical Properties of SASE Radiation

We have seen in the previous section that the transverse coherence of SASE FEL radiation is very good, which is extremely important for structural analysis experiments. The longitudinal coherence, however, is rather poor. The start-up from shot noise is the reason why SASE radiation exhibits strong shot-to-shot fluctuations in the wavelength spectrum and in the FEL pulse energies. This will be analyzed now.

7.3.1 Simple Model of a SASE Pulse

Truly monochromatic light is an idealization and does not occur in nature as it corresponds to an infinitely long harmonic wavetrain. Any observable wavetrain has a finite duration and therefore subtends a certain range of wavelengths. The radiation is called *quasi-monochromatic* if the bandwidth is much smaller than the central wavelength. This is usually the case for conventional laser radiation. (An exception are lasers with ultrashort pulses such as femtosecond titanium-sapphire lasers. Here the central wavelength is 800 nm and the bandwidth may be as large as 100 nm). An important feature of quasi-monochromatic radiation is that there exist fixed phase relations among the different wavelength components.

In contrast to this, SASE FEL radiation is in general of a chaotic nature since it originates from shot noise in the electron beam (see Sect. 10.5.1). The SASE pulses are composed of several sequential wavetrains with arbitrary phase shifts in between. An illustration is given in Fig. 7.9. Within a given SASE pulse, the phase relations are fixed, but the phase shifts fluctuate from pulse to pulse.

What is the consequence for the emitted SASE spectrum? We study this in a model calculation with three sequential wavetrains of equal wavelength, each comprising 100 periods. The three Fourier transforms have to be added taking into account the proper phase factors. The spectral intensity of the SASE pulse is proportional to the absolute square of this sum and consists of two terms: (1) the sum of the spectral intensities of the three wavetrains, and (2) the interference terms among the trains. The interference terms drop out when computing the average spectrum of many FEL pulses. This average spectrum has a smooth lineshape, indicated by the dashed blue curves in Fig. 7.10, and its bandwidth of about 1 % is determined by the 100 periods in a single wavetrain. In contrast to this, the single-shot spectra of individual SASE pulses (red curves in Fig. 7.10) exhibit a considerable fine structure caused by interference: they are characterized by narrow spikes which fluctuate in position and peak height. The spike width is about 0.3 % because the overall wavetrain comprises 300 periods.

The characteristic shot-to-shot fluctuations have been experimentally verified. Two measurements at different wavelengths and different electron bunch lengths are shown in Fig. 7.11. The SASE pulses at a wavelength of 97 nm were produced by short electron bunches while those at 14.6 nm were produced by long electron

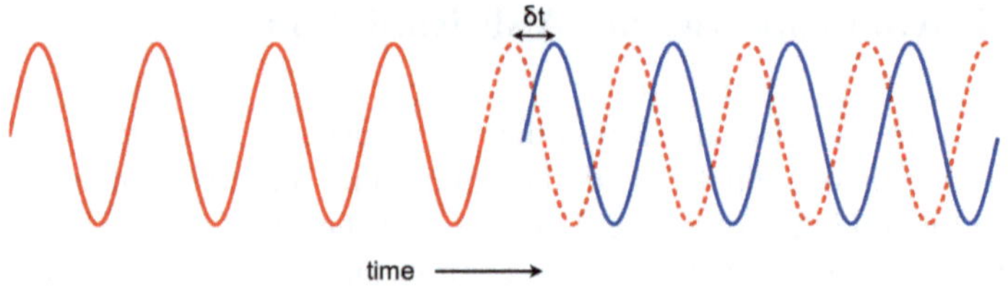

Fig. 7.9 Very simple model of a SASE pulse: a sequence of several harmonic wavetrains with arbitrary phase shifts in between. These phase shifts fluctuate from pulse to pulse. For clarity, only two wavetrains are shown.

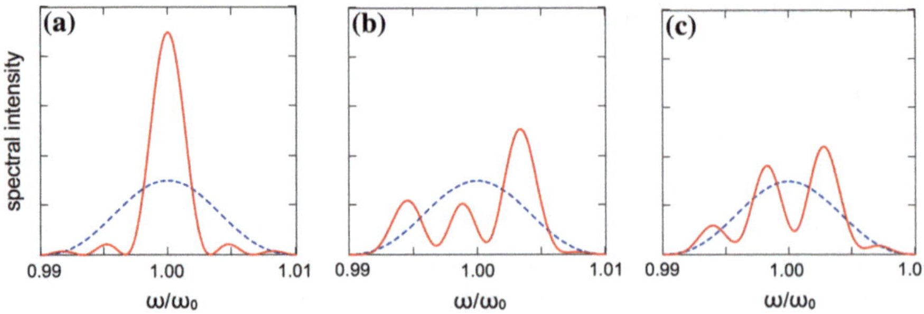

Fig. 7.10 Averaged SASE spectrum (*dashed blue curves*) and single-shot spectra of individual SASE pulses (*red curves*). **a** All phase shifts are zero, **b** and **c** randomly chosen phase shifts.

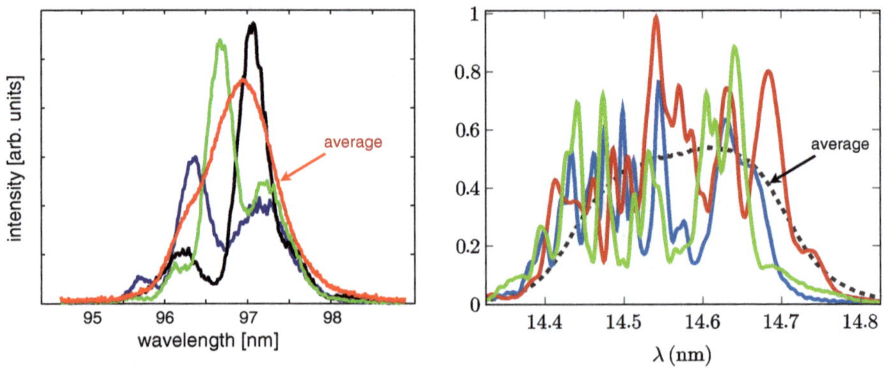

Fig. 7.11 *Left*: The measured spectra of individual SASE FEL pulses at an average wavelength of 97 nm [11, 12]. The single-shot spectra show two to three spikes which fluctuate in size and position from shot to shot. The average spectrum of 100 FEL pulses is wider than the individual spikes, as expected from our model calculation. *Right*: Spectra of individual SASE pulses and the averaged spectrum at 14.6 nm for long electron bunches [20]. The average number of spikes in the single-shot spectra is about 12.

bunches. The fine structure of SASE pulses was quite different in the two cases, the average number of spikes being 2.6 for the short electron bunches and 12 for the long bunches.

7.3.2 Coherence Time

An important quantity for the characterization of chaotic light is the coherence time, i.e. the time over which there exists a definite phase relation in the field. We want to study the coherence properties of the SASE FEL light emerging from an undulator with an active length of $L_u \approx 16\,L_{g0}$, where according to Fig. 5.9 the microbunching is very pronounced. The complex electric field amplitude at $z = L_u$ is denoted by $\mathcal{E}(t)$. The first-order correlation function is defined by the expression [21]

$$\mathcal{C}(\tau) = \frac{\langle \int \mathcal{E}(t)\mathcal{E}^*(t+\tau)\,dt\rangle}{\langle \int |\mathcal{E}(t)|^2\,dt\rangle} \tag{7.12}$$

in which the time integration extends over one SASE pulse and the brackets denote the sample average over all observed SASE pulses. Now we carry out a Fourier transformation

$$\mathcal{E}(t) = \frac{1}{2\pi}\int \tilde{\mathcal{E}}(\omega)\exp(-i\omega t)d\omega$$

and find

$$\mathcal{C}(\tau) = \frac{\left\langle \int \left|\tilde{\mathcal{E}}(\omega)\right|^2 \exp(i\,\omega\,\tau)d\omega\right\rangle}{\left\langle \int \left|\tilde{\mathcal{E}}(\omega)\right|^2 d\omega\right\rangle}. \tag{7.13}$$

Proof:

$$\int \mathcal{E}(t)\mathcal{E}^*(t+\tau)\,dt \propto \iiint \tilde{\mathcal{E}}(\omega)\tilde{\mathcal{E}}^*(\omega')\exp(-i\omega t)\exp(i\omega'(t+\tau))dt\,d\omega\,d\omega'$$
$$\propto \int \left|\tilde{\mathcal{E}}(\omega)\right|^2 \exp(i\,\omega\,\tau)d\omega$$

since $\int \exp(i(\omega' - \omega)t)\,dt = 2\pi\,\delta(\omega' - \omega)$.

The Fourier component of the field is proportional to the product of the transmission function $H(\omega, z)$ and the spectral current density $J(\omega)$, see Eq. (7.7). This current density results from shot noise in the electron beam, and since shot noise has a white spectrum, $J(\omega)$ is independent of frequency (compare Eq. (10.115) in Sect. 10.5.1). Using the Gaussian approximation (7.9) of the transmission function we obtain

$$\tilde{\mathcal{E}}(\omega) \propto \exp\left(-\frac{(\omega-\omega_r)^2}{4\sigma_\omega^2}\right) \;\Rightarrow\; \mathcal{E}(t) \propto \exp\left(-\frac{t^2}{4\sigma_t^2}\right) \text{ with } \sigma_t = \frac{1}{2\sigma_\omega}.$$

From Eq. (7.13) follows then that the correlation function is also a Gaussian

$$\mathcal{C}(\tau) = \exp\left(-\frac{\sigma_\omega^2 \tau^2}{2}\right). \tag{7.14}$$

Following [22, 23] we define the coherence time by

$$\boxed{\tau_{\text{coh}} = \int_{-\infty}^{\infty} (\mathcal{C}(\tau))^2 \, d\tau = \frac{\sqrt{\pi}}{\sigma_\omega} = 2\sqrt{\pi}\,\sigma_t \,.} \tag{7.15}$$

Remember that σ_ω depends on the distance z traveled in the undulator, see Eq. (5.14).

7.3.3 Pulse Energy Fluctuations

Consider now a "flat-top" bunch with a time duration of T_{bunch}. The average number of wave packets (spikes) in the wavelength spectra is

$$M = \frac{T_{\text{bunch}}}{\tau_{\text{coh}}} \tag{7.16}$$

if the bunch duration exceeds the coherence time because each wave packet has a time duration of about τ_{coh} and thus $M = T_{\text{bunch}}/\tau_{\text{coh}}$ non-overlapping wave packets can exist in the bunch. For $T_{\text{bunch}} \leq \tau_{\text{coh}}$ only one wave packet can exist and hence $M = 1$. The wave packets are sometimes called the "longitudinal modes". The frequency width of the individual spikes in the spectrum is equal to the Fourier transform limit determined by the bunch duration

$$\Delta\omega_{\text{spike}} = \frac{2\sqrt{2\ln 2}}{T_{\text{bunch}}} \,. \tag{7.17}$$

This formula holds for a bunch with a flat time profile.

In the exponential growth regime the radiation pulse energy U_{rad} fluctuates according to the so-called gamma distribution, see Sect. 10.5.2:

$$p_M(u)du = \frac{M^M \, u^{M-1}}{\Gamma(M)} \exp(-M\,u)du \quad \text{with} \quad u = \frac{U_{\text{rad}}}{\langle U_{\text{rad}} \rangle} \,. \tag{7.18}$$

Here U_{rad} is the energy of an individual radiation pulse, $\langle U_{\text{rad}} \rangle$ the average energy of many pulses, $u = U_{\text{rad}}/\langle U_{\text{rad}} \rangle$ the normalized pulse energy, and $\Gamma(M)$ is the gamma function of argument M. The mean value and the variance of the normalized FEL pulse energy are

$$\langle u \rangle = \int_0^\infty u \; p_M(u)du = 1, \quad \sigma_u^2 = \left\langle (u - \langle u \rangle)^2 \right\rangle = \frac{1}{M}. \tag{7.19}$$

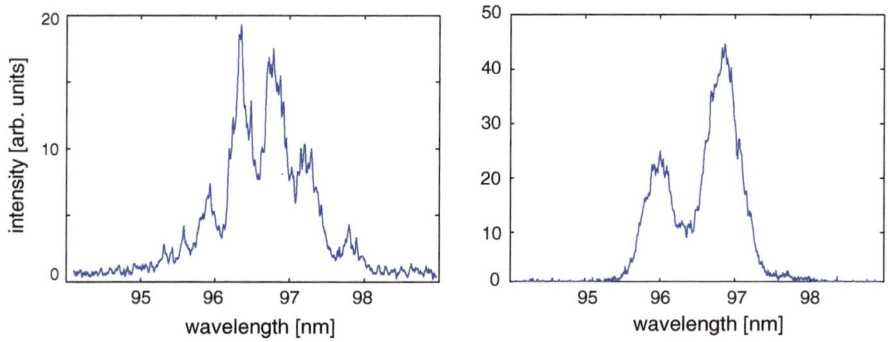

Fig. 7.12 *Left*: Single-shot spectrum of a long FEL pulse. The average number of modes is $M = 6$. *Right*: Single-shot spectrum of a short FEL pulse. Here the average number of modes is $M = 2.6$.

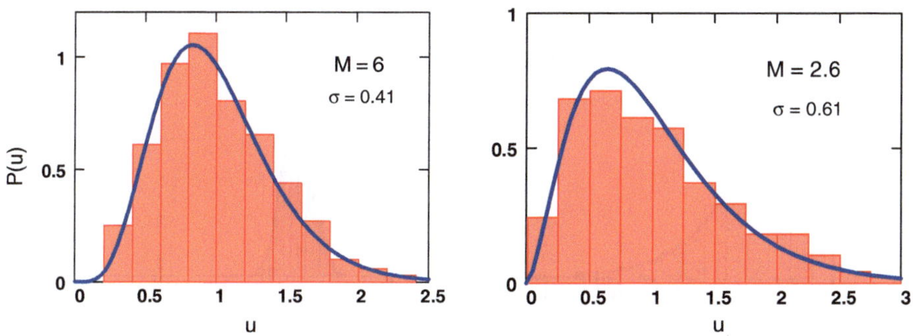

Fig. 7.13 Fluctuation of SASE pulse energy for long and short electron bunches [12]. The mode number is $M = 6$ for the long pulse and the variance is $\sigma_u = 41\,\%$. For the short pulse one finds $M = 2.6$ and $\sigma_u = 61\,\%$. The data have been taken in the exponential gain regime. *Solid curves* gamma distribution (7.18).

As mentioned above, the number of modes is equivalent to the number of spikes seen in the wavelength spectra.

As a nice illustration we show in Figs. 7.12 and 7.13 data obtained at a wavelength of 97 nm [12]. In a long FEL pulse more wave packets are excited than in a short pulse, and the probability distribution of SASE pulse energy has a correspondingly lower variance. This is indeed verified by the data. When one selects a single longitudinal mode ($M = 1$) by means of a monochromator the probability distribution changes dramatically into a negative-exponential distribution, see Fig. 7.14. We note that the pulse energy fluctuations are large in the exponential gain regime and reach 60–70 %. This has also been verified in a measurement at a wavelength of 13.7 nm (Fig. 7.15). From the experimental value $M = 1.9$ and an estimated coherence time of $\tau_{coh} \approx 4.2 \pm 0.5$ fs one obtains an estimate for the radiation pulse length at the end of the exponential regime of about 8 ± 1 fs [13].

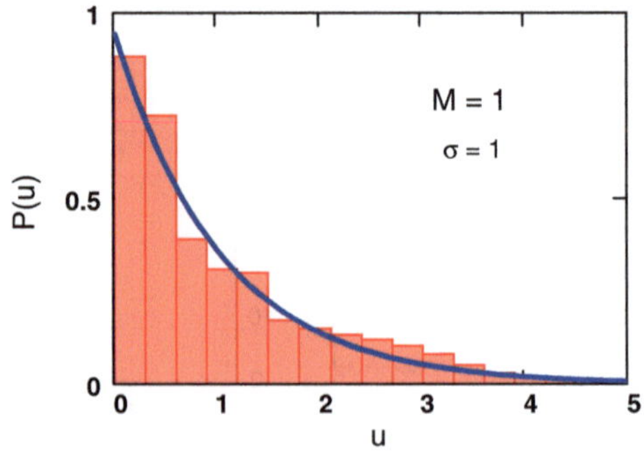

Fig. 7.14 Pulse energy distribution in FEL radiation having passed a monochromator slit. Here one has $M = 1$ and $\sigma_u = 100\,\%$.

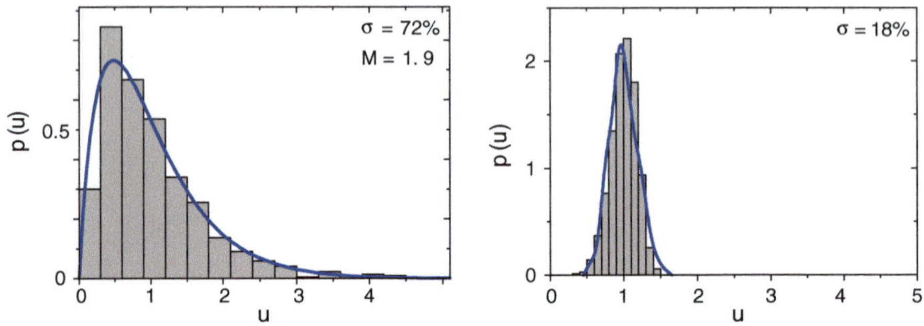

Fig. 7.15 *Left*: Measured probability distribution of FEL pulse energy in the exponential regime at a wavelength of 13.7 nm [13]. *Solid curve* gamma distribution (7.18). The average pulse energy is 1 μJ. *Right*: Probability distribution of FEL pulse energy in the saturation regime. The average pulse energy is 40 μJ. Here the curve has been computed with the code FAST.

When saturation is reached the gamma distribution is no longer applicable and the fluctuations drop to less than 20 %. The regimes of exponential gain and saturation are compared in Fig. 7.15.

A thorough characterization of the 840 nm SASE radiation produced at VISA can be found in Ref. [24]. A short gain length of less than 18 cm has been obtained and a gain of 2×10^8 at saturation. For uncompressed bunches the SASE radiation spectra exhibit 4–5 longitudinal modes, and the shot-to-shot intensity fluctuations are described by a gamma distribution with $M = 4.3$, while for compressed bunches only one mode is seen and the observed fluctuations are well fitted with the negative exponential distribution (i.e. the gamma distribution with $M = 1$). In another experiment at VISA [25] the microbunching was verified by detecting coherent optical transition radiation at the first and second harmonic of the SASE FEL radiation (at 845 and 422 nm).

The spikes in the wavelength spectra have their origin the presence of several wave packets that are separated in time. Consequently, the FEL pulses in time-domain will

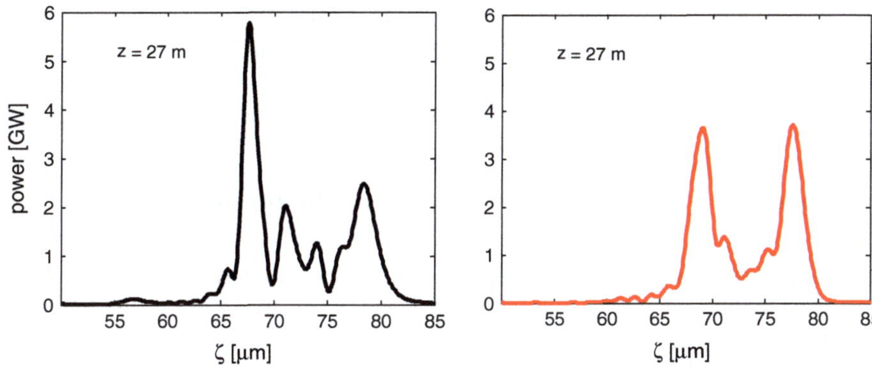

Fig. 7.16 The computed time structure of two different FEL pulses in FLASH at the end of the 27 m long undulator. We thank Igor Zagorodnov for carrying out the GENESIS calculation.

also feature sharp spikes. All FEL simulation codes predict indeed the existence of sharp spikes in the time-domain. As an example we show in Fig. 7.16 a computation made with the code GENESIS.

7.4 FEL Seeding with Coherent Radiation

Self-amplified spontaneous emission originates from spontaneous undulator radiation and is hence a stochastic process. The start-up from shot noise is the reason why SASE FEL radiation exhibits strong shot-to-shot fluctuations in the wavelength spectrum, as discussed in Sect. 7.3. The pulse energy fluctuations are strong in the exponential gain regime but drop considerably when FEL saturation is approached. A long bunch can be subdivided into sections that are each one coherence length long. In the SASE process the microbunching takes place independently in different sections which then radiate incoherently. Therefore the longitudinal coherence of SASE radiation is poor compared to conventional laser light.

The straightforward solution to overcome these limitations is to initiate the FEL process by a coherent source. Seeding with external laser-like radiation induces uniform microbunching all along the electron bunch and leads to the emission of FEL radiation which is fully coherent both in space and time, and which is free from large pulse-energy fluctuations, provided the seeding source is sufficiently stable. The much better FEL beam quality is very important, not only for spectroscopic experiments, but also for a whole class of time-domain experiments requiring a well-known and controllable optical phase.

External seeding of an extreme ultraviolet (EUV) or soft X-ray FEL requires an intense laser-like source. At these short wavelengths, conventional lasers with sufficient power to dominate the competing SASE process are not available today. Various seeding methods have been proposed three of which will be described below: generation of high-order harmonics of an optical or infrared laser through nonlinear processes in a noble gas, high-gain harmonic generation, and echo-enabled harmonic generation. A fourth method, self-seeding, is well suited for the hard X-ray regime where the other schemes become inefficient. Self-seeding will be discussed in Sect. 9.2.3.

7.4.1 Seeding by Higher Harmonics of an Infrared Laser

High-order laser harmonics generated in a gas open the way to FEL seeding at short wavelengths. A successful experiment has been carried out at the SPring-8 Compact SASE Source (SCSS) test accelerator [26]. The test accelerator is depicted in Fig. 7.17. A pulsed titanium:sapphire (Ti:Sa) laser (wavelength 800 nm, pulse duration 100 fs FWHM) is focused on a xenon gas cell where it generates odd higher harmonics. The fifth harmonic with a wavelength of 160 nm, a pulse energy close to 1 μJ and duration of 50 fs (FWHM) is selected for seeding. The 160 nm seed beam is spatially and temporally overlapped with a 150 MeV electron beam in two consecutive undulator sections.

Figure 7.18 compares typical spectral intensity distributions of the seeded FEL emission, of the unseeded emission (seed beam switched off) and of the seed beam itself. The data were obtained with the first undulator section tuned to 160 nm, while the second undulator section was off resonance. At the fundamental

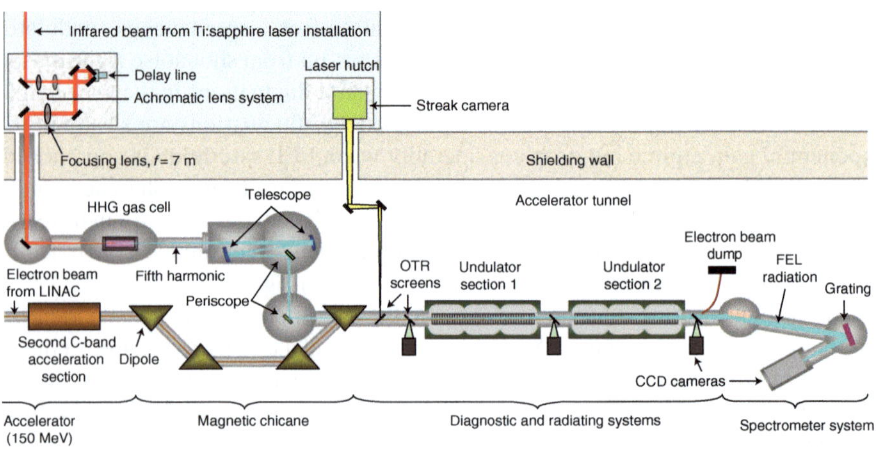

Fig. 7.17 Layout of the SSCS test accelerator and seeding experiment. A powerful Ti:Sa laser with a fundamental wavelength of 800 nm is loosely focused on a xenon gas cell where it generates higher harmonics. Using the telescope and periscope optics (CaF_2 mirrors), the fifth harmonic is spectrally selected. The 160 nm seed beam is overlapped with the 150 MeV electron beam in two consecutive undulator sections. The relative alignment of electron and photon beam is monitored on optical transition radiation (OTR) screens. The FEL radiation is spectrally analyzed using a diffraction grating and a CCD camera. Reprinted by permission from [26]. © 2008 by Macmillan Publishers Limited.

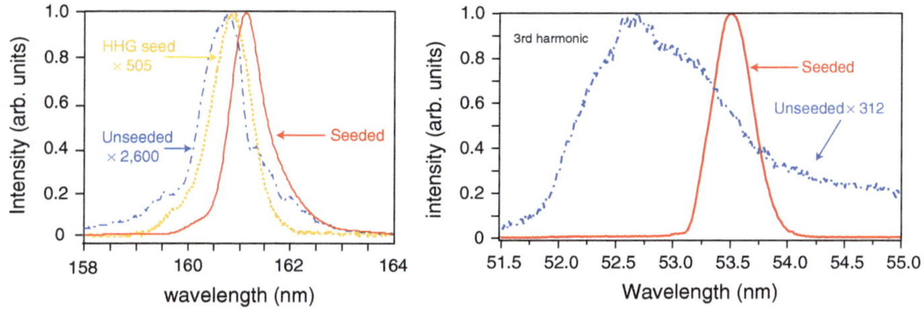

Fig. 7.18 *Left*: The spectral distributions of the seeded FEL radiation (*red curve*), the unseeded FEL radiation, multiplied with a scale factor of 2600 (*dashed blue curve*), and the seed beam multiplied with a factor of 505 (*dotted yellow curve*) at the fundamental wavelength (160 nm). *Right*: Spectra of seeded FEL radiation (*red curve*) and unseeded FEL radiation, multiplied with a factor of 312 (*dashed blue curve*) at the third-harmonic wavelength of 53.5 nm. Reprinted by permission from [26]. © 2008 by Macmillan Publishers Limited.

wavelength of 160 nm, the seeded FEL intensity is a factor of 500 larger than the seed beam intensity and even a factor of 2600 larger than the unseeded radiation. Odd higher harmonics are also seen. At the third harmonic ($\lambda = 53$ nm) the seeded FEL intensity exceeds the unseeded FEL intensity by a factor of 312 and at the fifth harmonic ($\lambda = 32$ nm) by a factor of 47.

In a later experiment at SCSS [27] the 13th harmonic of a Ti:Sa laser was used as a seeding source for a 250 MeV electron beam. The radiation intensity at the seed wavelength of 61.2 nm increased by a factor of 650 compared to the unseeded case. Seeding with a wavelength of 38 nm has recently been demonstrated at FLASH [28].

7.4.2 High-Gain Harmonic Generation

FEL light can be produced by the higher harmonics of external seed radiation via the so-called High-Gain Harmonic Generation (HGHG) process [29]. The principle is explained in Fig. 7.19. The electron beam and a powerful laser beam or laser-like beam of wavelength λ_1 are sent together through a short undulator magnet, called the *modulator*, which is tuned to λ_1. The energy of the electron beam is periodically modulated by the interaction with the seed beam. A precondition is a seed beam of sufficient power to generate an energy modulation well above the energy spread of the electron beam. A magnetic chicane converts the energy modulation into a charge density modulation. Plotted as a function of the internal bunch coordinate ζ, the charge density exhibits a periodic sequence of narrow spikes with a spacing equal to the seed wavelength λ_1, see graph (c) in Fig. 7.19. The Fourier expansion of $\rho(\zeta)$ contains higher harmonics with angular frequencies $\omega_n = n\,\omega_1$ and wavelengths $\lambda_n = \lambda_1/n$. The long undulator, called the *radiator*, is tuned to one of the higher-harmonic wavelengths, and in the radiator the density-modulated beam produces FEL radiation at this wavelength.

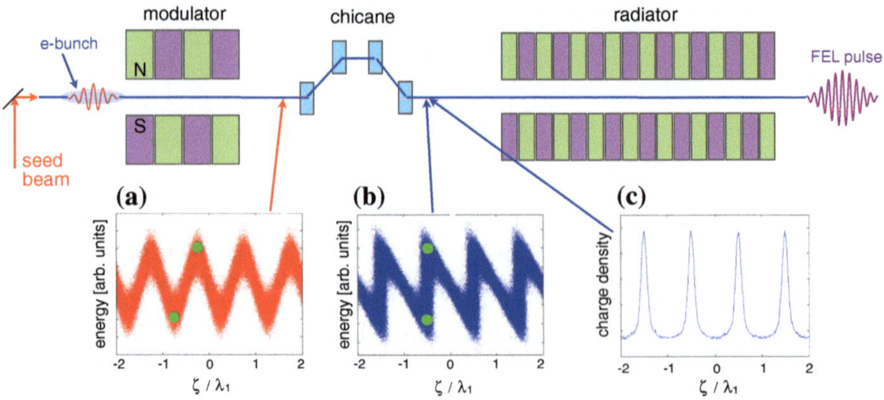

Fig. 7.19 Principle of the high-gain harmonic generation process. The *top graph* shows the experimental setup. The longitudinal phase space distributions (courtesy of D. Xiang) are displayed in *the bottom graphs*. The energy distribution in the bunch is plotted as a function of ζ/λ_1. **a** Downstream of the modulator, **b** downstream of the magnetic chicane. Graph **c** shows the microbunching: charge density as a function of ζ/λ_1.

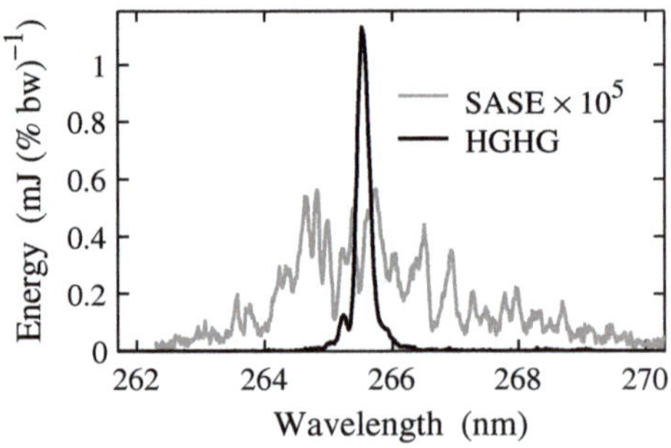

Fig. 7.20 Single shot HGHG spectrum obtained at the DUV FEL for a 30 MW seed beam. The *gray line* is the single shot SASE spectrum far below saturation, measured without seed beam. Reprinted with permission from [29], © 2003 by the American Physical Society.

The first experimental results on high-gain harmonic-generation in the ultraviolet were obtained with the deep ultraviolet free-electron laser (DUV FEL) at Brookhaven National Laboratory [29]. A Ti:Sa laser provided the seed beam with $\lambda_1 = 800$ nm for generating third-harmonic radiation at 266 nm. The results (Fig. 7.20) confirmed the predictions for HGHG FEL operation: stable wavelength spectrum, narrow bandwidth, and small pulse-energy fluctuation.

The free-electron laser FERMI at the synchrotron radiation laboratory Elettra in Trieste, Italy is a user facility based on the HGHG principle. In the initial experiments at FERMI, reported in [30], the third harmonic of a titanium-sapphire laser has been used for seeding. The 260 nm seed pulses have a length of 150 fs (FWHM), a peak power of about 100 MW and a bandwidth of 0.8 nm. The modulator has a period of 100 mm and a length of 3 m. The radiator is composed of six 2.4 m long undulator sections with a period of 55 mm. A special feature is the possibility to switch between linear and circular polarization.

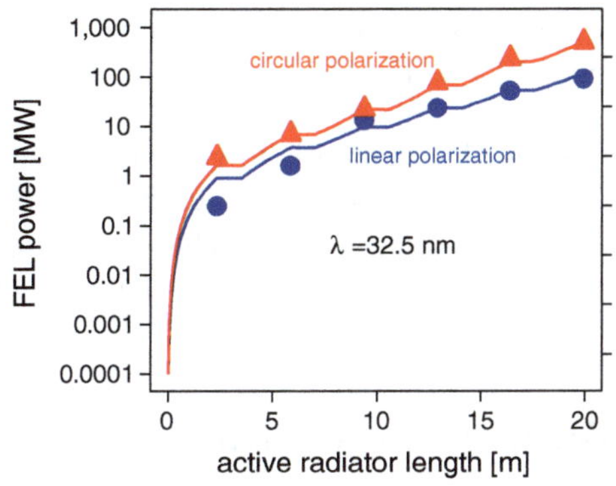

Fig. 7.21 Measured FEL power at 32.5 nm as a function of the active radiator length. Data are shown for both circular polarization (*red symbols*) and linear polarization (*blue symbols*). Adapted with permission from [30]. © 2012 by Macmillan Publishers Limited.

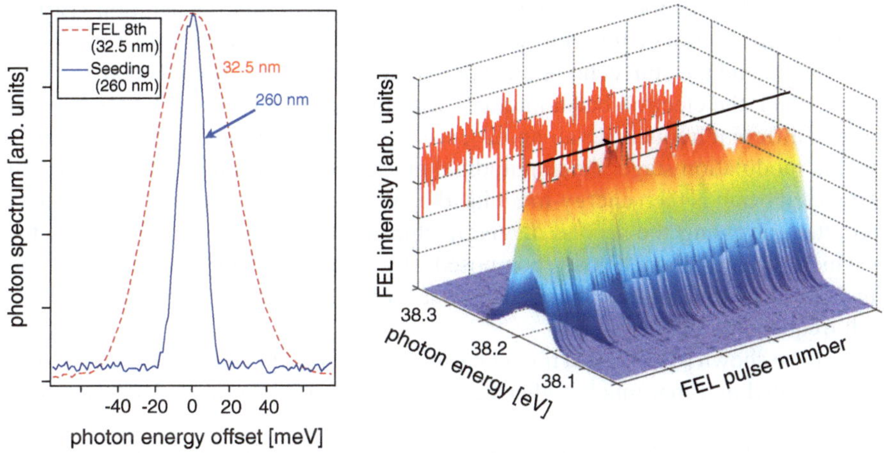

Fig. 7.22 Single-shot and multi-shot spectra at 32.5 nm. *Left*: Measured FEL and seed laser spectrum (*dashed red* and continuous *blue lines* respectively). *Right*: Acquisition of 500 consecutive FEL spectra. Adapted with permission from [30], © 2012 by Macmillan Publishers Limited.

The power gain curve of the FEL radiation produced at the 8th harmonic (32.5 nm) is shown in Fig. 7.21. The power gain length is 2.5 m for planar polarization and 2.0 m for circular polarization. As theoretically expected (see Sect. 4.9), a shorter gain length and a higher output power are obtained in the case of circular polarization, owing to a better coupling between electromagnetic field and electron beam. The FERMI FEL achieves saturation with a radiator of 6–8 gain lengths, while for a SASE FEL about 20 gain lengths are needed.

An attractive feature of seeded FELs is the monochromatic spectrum. The results obtained at FERMI are shown in Fig. 7.22, where both a single-shot spectrum and a sequence of 500 consecutive single-shot spectra are displayed. The seeded-FEL spectrum consists of a single spectral line with a relative width of 0.05 % while the SASE bandwidth is about 0.13 %. The FEL pulse energy fluctuations are in the 10 % range.

The basic idea of the HGHG method is to generate a microbunching with the periodicity λ_l of the seed radiation but with microbunches that are much shorter than λ_l so that the Fourier spectrum of the bunch contains higher harmonics. The efficiency of HGHG drops rapidly with increasing harmonic number, mainly because of beam energy spread.

7.4.3 Echo-Enabled Harmonic Generation

The HGHG-microbunches have a Gaussian-like charge density profile and do not posses a substructure. In [31] a novel method was proposed for generating microbunches with a considerable substructure, thus opening the way to much higher harmonics. The method was called Echo-Enabled Harmonic Generation (EEHG) because there is some similarity with the well-known beam echo effect in circular accelerators and the spin echo in nuclear magnetic resonance spectroscopy.

In EEHG, the electron beam interacts with two laser pulses which may have the same or different frequencies ω_1 and ω_2, see Fig. 7.23. After the magnetic chicane of the first modulation stage, the longitudinal phase space is strongly filamented. This filamented beam is then energy-modulated by the second laser beam in modulator 2, and after traversing the second magnetic chicane it has acquired a microstructure in which the microbunches are composed of a number of very narrow peaks, see Fig. 7.23. It is obvious that such a bunch has a Fourier spectrum extending to very high frequencies.

It can be shown [31, 32] that higher harmonics with frequencies $\omega = n_1\omega_1 + n_2\omega_2$ are generated and that much higher harmonic numbers than in the HGHG process can be achieved.

A proof-of-principle experiment was carried out using laser wavelengths of 795 and 1590 nm [33]. Spectral lines at 370, 405 and 465 nm were observed whose frequencies correspond to linear combinations of ω_1 and ω_2.

Echo-enabled harmonic generation in the ultraviolet has been observed at the Shanghai Deep Ultraviolet FEL (SDUV-FEL) facility [34]. A linac provides a 135 MeV electron beam with a bunch charge of 100–300 pC, a normalized emittance of 4 mm mrad, an energy spread of 0.2 %, and a pulse duration of 2–3 ps (FWHM) after compression. The two laser seeds come from a 1047 nm laser with 8.7 ps pulse length and a tunable pulse energy of up to 60 mJ. Energy modulations are applied in the modulators M1 (10 periods with a period of 65 mm and $K = 1.6$) and M2 (10 periods with a period of 50 mm and $K = 2.0$). The measurements were performed at the third harmonic of the laser seed. The electron beam was first sent through the radiator undulators to achieve optimized SASE. The laser seed was then introduced to interact with the electron beam in the second modulator (typical HGHG configuration). At this stage, a narrow-bandwidth HGHG signal was observed at 345 nm (Fig. 7.24). Finally, another synchronized laser seed was gradually guided to interact

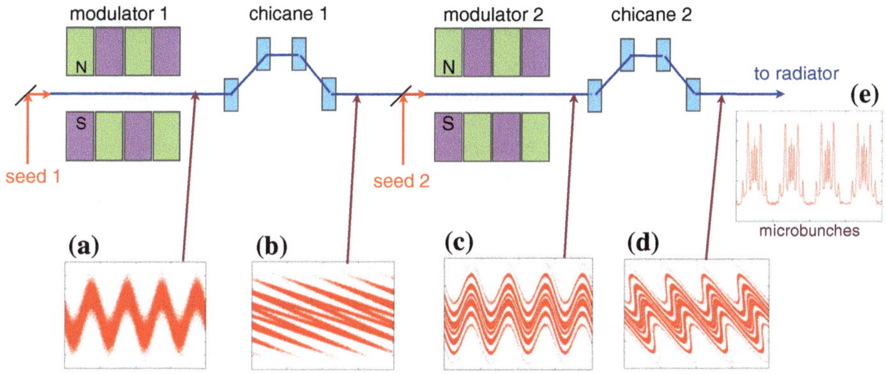

Fig. 7.23 *Top graph*: Schematic setup for echo-enabled harmonic generation. *Bottom graphs*: The longitudinal phase space distributions are displayed (courtesy of D. Xiang). The energy distribution in the bunch is plotted as a function of ζ/λ_1. **a** Downstream of modulator 1, **b** downstream of magnetic chicane 1, **c** downstream of modulator 2, **d** downstream of magnetic chicane 2. Graph **e** shows the microstructure in the outgoing bunch.

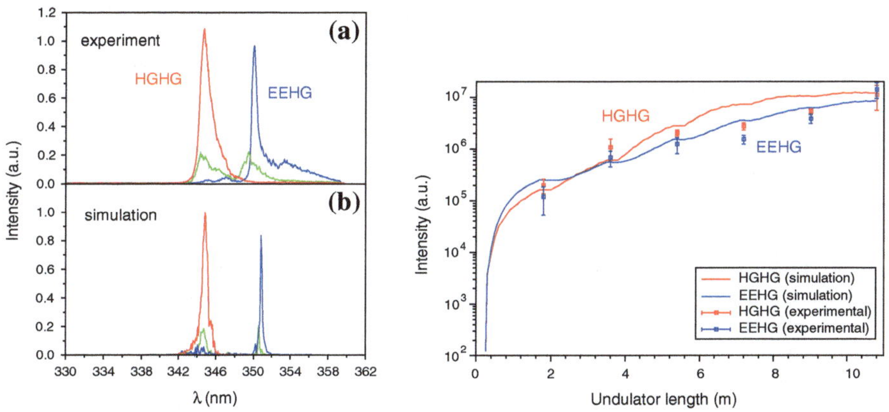

Fig. 7.24 *Left*: Spectra for FEL radiation. **a** Experimental results (*red line*, HGHG; *blue line*, EEHG). **b** Simulation results. *Right*: Gain curves for EEHG and HGHG operation. The intensity is plotted as a function of the active undulator (radiator) length. Reprinted with permission from [34]. © 2012 by Macmillan Publishers Limited.

with the electron beam in the first modulator to realize the EEHG seeding, resulting in the appearance of a narrow signal at a slightly longer wavelength (350 nm). The shift from 345 to 350 nm is caused by the different dependencies of the HGHG and EEHG wavelengths on the energy chirp of the electron beam and the dispersions in the magnetic chicanes, see Ref. [34].

References

1. A.M. Kondratenko, E.L. Saldin, Generation of coherent radiation by a relativistic electron beam in an ondulator. Part. Accel. **10**, 207 (1980)
2. R. Bonifacio, C. Pellegrini, L.M. Narducci, Collective instabilities and high-gain regime in a free electron laser. Opt. Comm. **50**, 373 (1984)
3. K.-J. Kim, Three-dimensional analysis of coherent amplification and self-amplified spontaneous emission in free-electron lasers Phys. Rev. Lett. **57**, 1871 (1986)
4. K.-J. Kim, An analysis of self-amplified spontaneous emission. Nucl. Instrum. Meth. A **250**, 396 (1986)
5. Z. Huang, K.-J. Kim, Review of x-ray free-electron laser theory. Phys. Rev. ST Accel. Beams **10**, 034801 (2007)
6. M. Hogan et al., Measurements of high gain and intensity fluctuations in a SASE free-electron laser. Phys. Rev. Lett. **80**, 289 (1998)
7. M. Hogan et al., Measurements of gain larger than 10^5 at 12 μm in a self amplified spontaneous-emission free-electron laser. Phys. Rev. Lett. **81**, 4867 (1998)
8. S.V. Milton et al., Exponential gain and saturation of a self-amplified spontaneous emission free-electron laser. Science **292**, 2037 (2001)
9. A.H. Lumpkin et al., Evidence for microbunching sidebands in a saturated free-electron laser using coherent optical transition radiation. Phys. Rev. Lett. **88**, 234801 (2002)
10. J. Andruszkow et al., First observation of self-amplified spontaneous emission in a free-electron laser at 109 nm wavelength. Phys. Rev. Lett. **85**, 3825 (2000)
11. V. Ayvazyan et al., Generation of GW radiation pulses from a VUV free-electron laser operating in the femtosecond regime. Phys. Rev. Lett. **88**, 104802 (2002)
12. V. Ayvazyan et al., A new powerful source for VUV radiation: demonstration of exponential growth and saturation at the TTF free-electron laser. Eur. Phys. J. D **20**, 149 (2002)
13. W. Ackermann et al., Operation of a free-electron laser from the extreme ultraviolet to the water window. Nat. Photonics **1**, 339 (2007)
14. V. Ayvazyan et al., First operation of a free-electron laser generating GW power radiation at 32 nm wavelength. Eur. Phys. J. D **37**, 297 (2006)
15. A. Tremaine et al., Experimental characterization of nonlinear harmonic radiation from a visible self-amplified spontaneous emission free-electron laser at saturation. Phys. Rev. Lett. **88**, 204801 (2002)
16. L. DiMauro et al., First SASE and seeded FEL lasing of the NSLS DUV FEL at 266 and 400 nm. Nucl. Instr. Meth. A **507**, 15 (2003)
17. T. Shintake, in *Status of Japanese XFEL project and SSCS test accelerator*, invited paper. FEL Conference, Berlin (2006)
18. R. Ischebeck, Transverse coherence of a VUV free electron laser, Ph. D. Thesis, University of Hamburg 2003, DESY-Thesis-2003-033
19. R. Ischebeck et al., Study of the transverse coherence at the TTF free electron laser. Nucl. Instr. Meth. A **507**, 175 (2003)
20. C. Behrens et al., Constraints on photon pulse duration from longitudinal electron beam diagnostics at a soft x-ray free-electron laser. Phys. Rev. ST Acc. Beams **15**, 030707 (2012)
21. K.-J. Kim, Z. Huang, R. Lindberg, in *Principles of Free Electron Lasers*. Lecture Notes, US Particle Accelerator School (to be published)
22. M.V. Klein, T.E. Furtak, *Optik* (Springer, Heidelberg, 1988)
23. E.L. Saldin, E.A. Schneidmiller, M.V. Yurkov, *The Physics of Free Electron Lasers* (Springer, Heidelberg, 2000)
24. A. Murokh et al., Properties of the ultrashort gain length, self-amplified spontaneous emission free-electron laser in the linear regime and saturation. Phys. Rev. E **67**, 066501 (2003)
25. A. Tremaine et al., Fundamental and harmonic microbunching in a high-gain self-amplified spontaneous-emission free-electron laser. Phys. Rev. E **66**, 036503 (2002)
26. G. Lambert et al., Injection of harmonics generated in gas in a free-electron laser providing intense and coherent extreme-ultraviolet light. Nat. Phys. **4**, 296 (2008)

27. T. Togashi et al., Extreme ultraviolet free electron laser seeded with high-order harmonic of Ti:sapphire laser. Opt. Express **19**, 317 (2011)
28. S. Ackermann et al., Observation of coherent 19 nm and 38 nm radiation at a free-electron laser directly seeded at 38 nm. Phys. Rev. Lett. **111**, 114801 (2013)
29. L.-H. Yu et al., First ultraviolet high-gain harmonic-generation free-electron laser. Phys. Rev. Lett. **91**, 074801 (2003)
30. E. Allaria et al., Highly coherent and stable pulses from the FERMI seeded free-electron laser in the extreme ultraviolet. Nat. Photonics **6**, 699 (Oct. 2012)
31. G. Stupakov, Using the beam-echo effect for generation of short-wavelength radiation. Phys. Rev. Lett. **102**, 074801 (2009)
32. D. Xiang, G. Stupakov, Echo-enabled harmonic generation free electron laser. Phys. Rev. ST Accel. Beams **12**, 030702 (2009)
33. D. Xiang et al., Demonstration of the echo-enabled harmonic generation technique for short-wavelength seeded free electron lasers. Phys. Rev. Lett. **105**, 114801 (2010)
34. Z.T. Zhao et al., First lasing of an echo-enabled harmonic generation free-electron laser. Nat. Photonics **6**, 360 (2012)

Chapter 8
The EUV and Soft X-Ray FEL in Hamburg

8.1 Introductory Remarks

The idea to use a long linear accelerator (linac) for providing the drive beam for an X-ray free-electron laser was conceived at the Stanford Linear Accelerator Center SLAC. In the Linac Coherent Light Source LCLS (see Chap. 9) a 1 km long section of the SLAC electron linac, which has been the major facility for elementary particle physics at Stanford since 1965, delivers the beam needed in the FEL. The SLAC linac is based on normal-conducting accelerating structures working at 2.856 GHz. The world's first linear collider SLC was realized utilizing this linac to accelerate electrons and positrons to 45.5 GeV and collide them after the traversal of an arc to study electro-weak physics at the Z^0 resonance.

For 25 years large groups of particle and accelerator physicists have been working on the development of linear electron-positron colliders in the TeV regime. While at Stanford and in Japan normal-conducting machines were designed, the TESLA collaboration decided for superconducting cavities. After a decade of intense research and development the collaboration succeeded in raising the accelerating field from a few Megavolts per meter to more than 35 MV/m in multi-cell niobium cavities [1]. The success of the TESLA cavity program was the essential motivation to base the future International Linear Collider ILC on the superconducting TESLA technology.

The TESLA Test Facility TTF was built at DESY with the intention to investigate the performance of superconducting cavities with an accelerated electron beam and to study whether the high beam quality needed in a collider could be achieved. Already at an early stage the decision had been taken to couple the envisaged TESLA collider with an X-ray FEL [2]. As a first test of this concept, the TTF machine was augmented with a 13.5 m long undulator magnet [3]. In 2000 the worldwide first vacuum-ultraviolet (VUV) free-electron laser began its operation at wavelengths between 80 and 180 nm. In the next step the energy of the linac was increased, a 27 m long undulator was installed and a hall for user experiments at five photon beam lines was added. In this configuration, FEL saturation was achieved for the first time in the extreme-ultraviolet (EUV) and soft X-ray regime at wavelengths between 30 nm and

P. Schmüser et al., *Free-Electron Lasers in the Ultraviolet and X-Ray Regime*, Springer Tracts in Modern Physics 258, DOI: 10.1007/978-3-319-04081-3_8,

4 nm. Results have been presented in Chap. 7. Since 2005 the FEL has been operating as a user facility named FLASH (Free-Electron Laser in Hamburg).

8.2 Layout of the Free-Electron Laser FLASH

The free-electron laser FLASH is shown schematically in Fig. 8.1. The electron bunches are produced in a laser-driven photo-injector and accelerated to energies up to 1250 MeV in a superconducting linac. The bunch charge is typically in the range of 0.1–0.5 nC. At intermediate energies of 150 MeV and 450 MeV the electron bunches are longitudinally compressed, thereby increasing the peak current from an initial value of some 30 A to more than 1000 A as required for the FEL operation. A third-harmonic cavity for an improved bunch compression system was installed in 2010.

The 27 m long planar undulator magnet consists of six segments. The field is generated by NdFeB permanent magnets and shaped by iron pole shoes [4]. The gap height is 12 mm, the period is $\lambda_u = 27.3$ mm, and the peak magnetic field is $B_0 = 0.48$ T. The field along the axis is almost purely sinusoidal:

$$B_y(z) = -B_0[\sin(k_u z) + b_3 \sin(3k_u z) + b_5 \sin(5k_u z) \ldots]$$

with the measured upper limits $|b_3| < 0.001$ and $|b_5| < 0.0005$. The deviation of the electrons from the ideal orbit is less than 10 μm. The good overlap between electron beam and light wave is a prerequisite for achieving a high gain in the lasing process. In the drift spaces between the segments, quadrupoles for beam focusing are installed as well as beam diagnostic instruments. Finally, a dipole magnet deflects the electron beam into a dump, while the FEL radiation propagates to the experimental hall.

The high-gain FEL process demands a bunched electron beam of extremely high quality which can be produced by a linac but not by a circular accelerator. Specifically, high peak current, low emittance, small energy spread and short bunch length are required. The technological challenges associated with these requirements and the technical solutions found are described in the following.

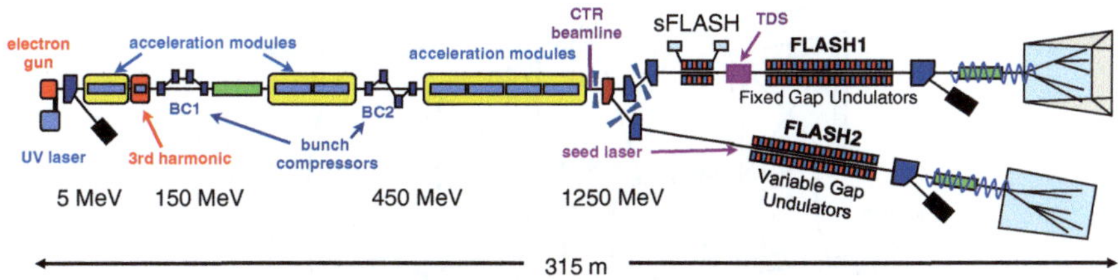

Fig. 8.1 Schematic view of the extreme-ultraviolet (EUV) and soft X-ray free-electron laser FLASH in Hamburg. The electron bunches are generated in a laser-driven photocathode. Seven acceleration modules are installed each containing eight superconducting cavities. Two magnetic chicanes BC1 and BC2 are used for longitudinal bunch compression. The section labelled sFLASH is dedicated for FEL seeding in the EUV and contains a 10 m long variable-gap undulator. The FEL beamline FLASH1 is equipped with a 27 m long fixed-gap undulator. A second FEL beamline (FLASH2) is in preparation which can be operated simultaneously with FLASH1 by sharing the many bunches accelerated in the superconducting linac. A variable-gap undulator permits the choice of a different wavelength.

8.3 Electron Source

The high bunch charge needed in a high-gain FEL can be accomplished with photocathodes which are illuminated with short ultraviolet laser pulses [5]. The electron source at FLASH, shown in Fig. 8.2, consists of a laser-driven photocathode which is mounted inside a $1\frac{1}{2}$-cell radio-frequency (RF) cavity. The cathode is made from molybdenum and is coated with a thin Cs_2Te layer to achieve a quantum efficiency for photoelectric emission of typically 5 %. The UV laser pulses are generated in a mode-locked solid-state laser system (Nd:YLF) built by the Max Born Institut, Berlin [6]. Another difference to conventional cathodes is the rapid acceleration to relativistic energies. The photocathode is located at the backplane of the half-cell where the accelerating field assumes its peak value of up to 60 MV/m. A static magnetic solenoid field is superimposed and provides transverse focusing in order to preserve a small beam cross section. The pulsed UV laser is synchronized to the 1.3 GHz radio-frequency of the linac with an rms precision of better than 60 fs.

It is not possible to generate the high peak current of several kA immediately in the gun because then huge space charge forces would arise and immediately disrupt the bunch. Therefore bunches with a modest current of some 30 A are produced by laser pulses of 10 ps duration, but even in this case the particles must be accelerated as quickly as ever possible to relativistic energies. In the relativistic regime the repulsive electric forces between the equal charges are largely canceled by the attractive magnetic forces between the parallel currents. Space charge forces are discussed in the next section.

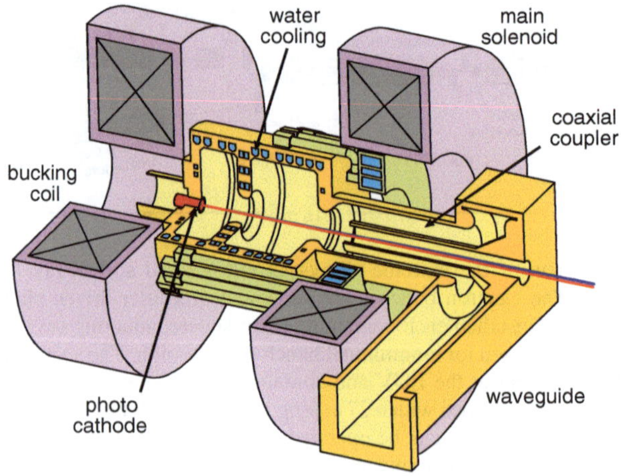

Fig. 8.2 Cut through the electron gun of FLASH. The Cs_2Te photocathode is mounted at the backplane of a 1.3 GHz $1\frac{1}{2}$-cell copper cavity. The cavity is excited in a TM_{010}-like mode, the electric field assumes its maximum value at the cathode. The RF power of about 5 MW is guided to the cavity through a wave guide and a coaxial coupler with minimum distortion of field symmetry. The UV laser beam is reflected onto the cathode by a small mirror outside the electron beam axis. A solenoid coil provides transverse focusing (see Sect. 8.4.2). A second solenoid, called "bucking coil", compensates the magnetic field in the cathode region where the photo-electrons have very low energy. (Courtesy of E. Vogel).

8.4 Space Charge Effects

Space charge forces in the intense electron bunches of a high-gain FEL have a profound influence on the beam dynamics and constitute in fact one of the main performance limitations of an X-ray FEL. We consider first space charge forces in highly relativistic bunches because these are easy to understand, and address then the rather intricate effects that happen in the low-energy electron cloud close to the photocathode.

8.4.1 Electric and Magnetic Forces Inside a Relativistic Bunch

We consider for simplicity a bunch of N relativistic electrons which are uniformly distributed in a cylinder of radius r_b and length L_b (measured in the laboratory system). In a co-moving coordinate system, the electrons are at rest and we have a pure Coulomb field inside the bunch. The number of particles and the bunch radius remain invariant ($r_b^* = r_b$) when transforming from the laboratory system to the co-moving coordinate system, but the length is Lorentz-expanded to $L_b^* = \gamma L_b$. For $\gamma \gg 1$ the bunch length L_b^* will be very much larger than the radius, hence one can compute the radial electric field inside the bunch using the approximation of an infinitely long cylindrical charge distribution. The field of an infinitely long charged cylinder has only a radial component

$$E_r^*(r) = \frac{-N\,e}{2\pi\varepsilon_0 L_b^*}\cdot\frac{r}{r_b^2} \quad \text{for} \quad r \le r_b\,,$$
$$= \frac{-N\,e}{2\pi\varepsilon_0 L_b^*}\cdot\frac{1}{r} \quad \text{for} \quad r \ge r_b\,. \tag{8.1}$$

A magnetic field does not exist because the charges are at rest in the co-moving system. Now we transform the field (8.1) into the laboratory system. This yields a radial electric field and an azimuthal magnetic field:

$$E_r(r) = \gamma E_r^*(r) = \frac{-N\,e}{2\pi\varepsilon_0 L_b}\cdot\frac{r}{r_b^2}\,, \quad B_\phi(r) = \frac{v}{c^2}\,E_r(r) \quad \text{for} \quad r \le r_b\,, \tag{8.2}$$

where v is the speed of the electrons. Outside the bunch ($r > r_b$) the fields drop as $1/r$. A test electron inside the bunch experiences a force due to the radial electric field and the azimuthal magnetic field which is computed using the formula $\boldsymbol{F} = -e\,(\boldsymbol{E} + \boldsymbol{v} \times \boldsymbol{B})$. The electric force points radially away from the axis, the magnetic force points inwards but is a bit weaker, hence the overall force points outwards and is thus a defocusing force, given by the expression

$$F_r(r) = \frac{N\,e^2}{2\pi\varepsilon_0 L_b}\cdot\frac{r}{r_b^2}\cdot\left(1 - \frac{v^2}{c^2}\right) = \frac{N\,e^2}{2\pi\varepsilon_0 L_b}\cdot\frac{r}{r_b^2}\cdot\frac{1}{\gamma^2} \quad \text{for } r \le r_b\,. \tag{8.3}$$

The residual force is a factor of $1/\gamma^2$ smaller than the electric force alone and vanishes in the ultra-relativistic limit $\gamma \to \infty$. This "repulsive" total force may lead to a radial blowup of the bunch. The electrons leave the gun at relativistic speed but their Lorentz factor $\gamma \approx 10$ is far too low to preserve a small normalized emittance over a large distance. Therefore the first acceleration module, raising the energy to 150 MeV, is mounted directly behind the electron gun.

Recognizing that repulsive space charge forces remain an unavoidable problem we may ask the question whether it is possible to counteract these internal forces at least partially by applying an external focusing field. For the cylindrical electron bunch with constant charge density, described in the previous section, this is in fact possible. In such a bunch the total space charge force depends linearly on the displacement r from the axis, see Eq. (8.3). A magnetic lens whose field grows linearly with r can indeed exert a focusing force that balances the internal force and preserves the transverse beam size.

Now it is important to realize that the transverse intensity distribution in a standard laser beam is described by a Gaussian, see Sect. 10.3. The electron bunch generated in a photocathode by such an laser beam will also possess a Gaussian transverse density distribution. The residual repulsive force inside a Gaussian bunch is given by the expression

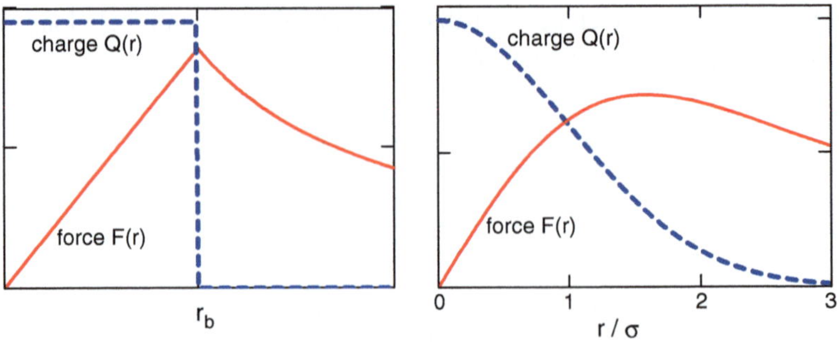

Fig. 8.3 *Left*: charge distribution $Q(r)$ and radial space charge force $F(r)$ in a cylindrical electron bunch with constant charge density and radius r_b. *Right*: charge distribution $Q(r)$ and radial space charge force $F(r)$ in a Gaussian bunch. The units for charge and force are arbitrary.

$$F_r(r) = \frac{N e^2}{2\pi\varepsilon_0 L_b r}\left[1 - \exp\left(-\frac{r^2}{2\sigma^2}\right)\right] \cdot \frac{1}{\gamma^2} . \tag{8.4}$$

The radial force inside a bunch with either a constant or a Gaussian charge distribution is depicted in Fig. 8.3. In the Gaussian bunch the force grows almost linearly with r for $0 \le r \le 0.8\,\sigma$ but then the slope dF/dr levels off, and for $r > 1.8\,\sigma$ the force even decreases. It is virtually impossible to compensate such a highly nonlinear force by means of an external magnetic field, but of course one can correct for part of it. For this reason the Gaussian charge distribution is quite undesirable in the electron injector region. Great effort is made to shape a UV laser beam with a flat transverse profile in order to produce cylindrical electron bunches with an approximately constant charge density. Also in the longitudinal direction a flat profile is desirable because otherwise the defocusing effects of the space charge forces would vary along the bunch while an external focusing force is necessarily the same all along the bunch.

8.4.2 Space Charge Forces in the Electron Gun

The electrons leave the photocathode with velocities much below c and form a charged cloud that is rapidly accelerated to relativistic speeds. The time-varying space charge fields in such an electron cloud cannot be computed analytically. Numerical calculations reveal that the radial force depends on the longitudinal position ζ inside the bunch; it is strongest in the center and falls off toward the head and tail of the bunch. The ζ dependence of the radial force distorts the phase space distribution and generates a fan-like structure, shown schematically in Fig. 8.4. The emittance within a short longitudinal slice, the so-called slice-emittance,[1] is not significantly affected, but the angular orientations of the phase space ellipses of different slices depend on their longitudinal position, with the consequence that the overall emittance is increased.

[1] We consider here the normalized emittance, see Sect. 8.9.

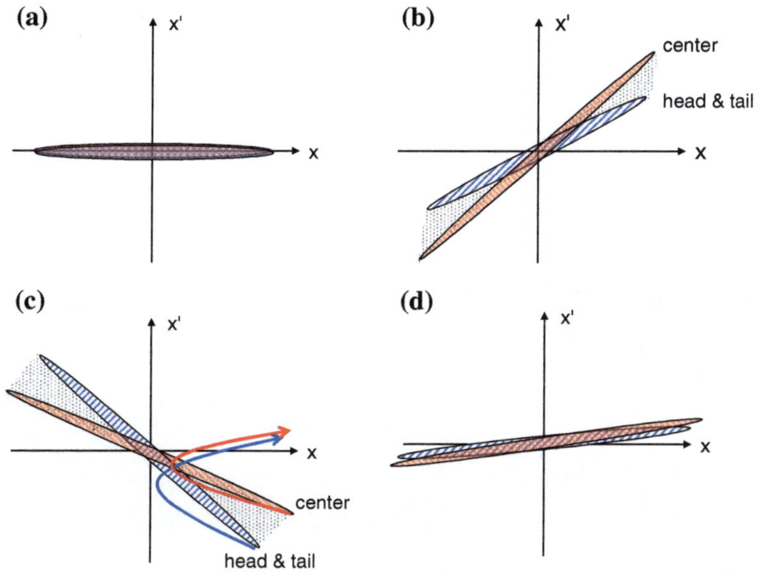

Fig. 8.4 Transverse phase space plots showing schematically the emittance growth due to space charge and its compensation by means of a focusing lens. **a** Initial phase space distribution near the photocathode. **b** Phase space distribution after beam transport to the lens. Not only the two ellipses, representing bunch slices at the center of the bunch and at the head or tail, are filled with particles but also the shaded area between them. This is because the strength of the space charge force varies continuously between center and head of the bunch. **c** Transformation of phase space distribution by a focusing lens. **d** Shrinkage of the fan-like structure in the drift and acceleration section behind the lens. We thank Klaus Flöttmann for useful discussions on this subject.

Fortunately, this radial blowup of the bunch can be counteracted by a clever focusing scheme known as *emittance compensation* [7]. (A more appropriate expression would be *emittance-growth compensation* because it is not the emittance as such that is compensated but rather its growth). With the help of a magnetic solenoid field the evolution of the fan-like phase-space structure can be reversed. The mathematical analysis [8] of the emittance-growth compensation process is beyond the scope of this book. Elaborate numerical computations using the code ASTRA [9] and thorough experimental studies [10] have been carried out to optimize the geometry of the solenoid coil arrangement and to determine the proper coil current.

8.5 Superconducting Linear Accelerator

The electron injector section is followed by seven 12.2 m long acceleration modules each containing eight superconducting niobium cavities. The cavities are made from pure niobium and consist of nine cells each. Figure 8.5 shows the layout of the nine-cell TESLA cavity [11]. The basic principles of superconducting cavities for particle acceleration are explained in [12]. An important property of superconductors is that

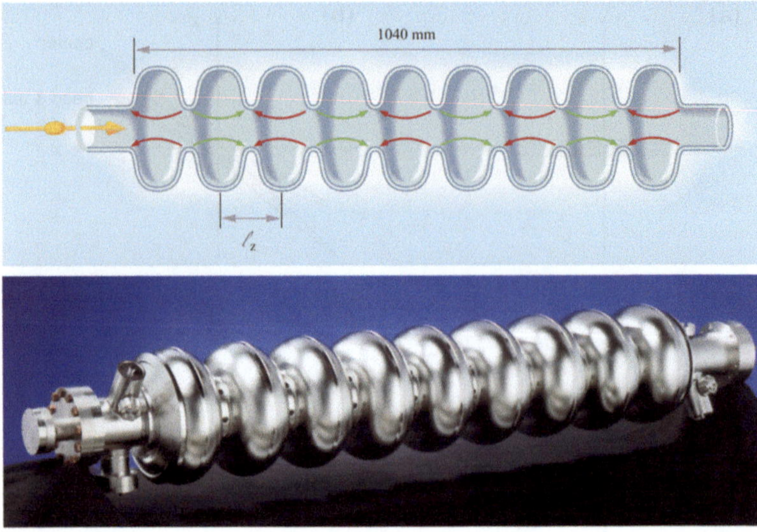

Fig. 8.5 Longitudinal cut and photo of the nine-cell superconducting cavity which is made from pure niobium and cooled by superfluid helium of 2 K. The resonance frequency is $f_0 = 1.3\,\text{GHz}$. The electric field lines are shown at the instant when an electron bunch has just entered the first cell. The length ℓ_c of a cell is chosen such that the field direction has inverted when the relativistic bunch has moved to the next cell. This is fulfilled for a cell length equal to half the RF wavelength, $\ell_c = c/(2f_0)$. Thereby it is ensured that the particles receive the same energy gain in each cell.

their resistance does not vanish in alternating electromagnetic fields, in contrast to the direct-current case. In a microwave cavity the oscillating magnetic field of the RF wave penetrates into the superconductor down to a depth of about 50 nm (London penetration depth) and induces forced oscillations of the "normal-conducting" electrons, namely those which are not bound in Cooper pairs. The resulting *microwave surface resistance* is many orders of magnitude smaller than that of normal copper cavities but is nevertheless responsible for significant RF power dissipation at the inner cavity surface. The dissipated power in an RF cavity is given by the formula

$$P_{\text{diss}} = \frac{R_{\text{surf}}}{2} H_{\text{RF}}^2 S\,, \tag{8.5}$$

where R_{surf} is the microwave surface resistance of the superconductor, H_{RF} is the magnitude of the RF magnetic field at the cavity surface, and S is the surface area. According to the Bardeen-Cooper-Schrieffer (BCS) theory of superconductivity the surface resistance depends exponentially on temperature, see [12] and the references quoted therein

$$R_{\text{BCS}} = \frac{A\,f_0^2}{T} \exp\left(-\frac{1.76\,T_c}{T}\right). \tag{8.6}$$

Here T_c is the critical temperature, and A is a coefficient that depends on the London penetration depth and other material properties. The surface resistance of a superconducting cavity can be written as

$$R_{\mathrm{surf}} = R_{\mathrm{BCS}} + R_{\mathrm{res}}$$

with a "residual resistance" of a few $\mathrm{n}\Omega$ that is caused by surface impurities. The exponential temperature dependence is the reason for cooling the high-field TESLA cavities with superfluid helium at 2 K ($R_{\mathrm{surf}} \approx 10\,\mathrm{n}\Omega = 10^{-8}\,\Omega$) instead of using pressurized normal liquid helium at 4.4 K ($R_{\mathrm{surf}} \approx 1000\,\mathrm{n}\Omega$).

The BCS surface resistance scales quadratically with the radio-frequency f_0, hence it is advantageous to build superconducting cavities with relatively low resonance frequencies. The value of 1.3 GHz chosen for TESLA is a good compromise between low surface resistance and manageable size of the cavities. The measured surface resistance of the high-purity niobium material used for the cavities follows the predicted exponential temperature dependence over a wide temperature range. Below 2 K the residual resistance of a few $\mathrm{n}\Omega$ begins to dominate.

The quality factor of a cavity can be expressed as the ratio of resonance frequency to the width of the resonance curve. It is inversely proportional to the surface resistance

$$Q_0 = \frac{f_0}{\Delta f} = \frac{G}{R_{\mathrm{surf}}} \,. \tag{8.7}$$

Here G is a "geometry factor" that depends only on the shape of the cavity but not on its material. A typical value for a one-cell cavity is $G = 300\,\Omega$. For niobium at 2 K the surface resistance is a few $\mathrm{n}\Omega$ so the quality factor is $Q_0 > 10^{10}$. In principle the quality factor should stay constant when the field in the cavity is raised from zero to an upper limit which is reached when the RF magnetic field approaches the critical magnetic field of the superconductor. For niobium at 2 Kelvin the critical field is $B_c \approx 200\,\mathrm{mT}$, corresponding to a maximum accelerating field $E_{\mathrm{acc}} \approx 45$ MV/m, averaged over the length of the cavity. In practice, however, the excitation curve $Q_0 = Q_0(E_{\mathrm{acc}})$ usually ends at a lower field due to "dirt effects" such as contamination of the inner cavity surface or field emission of electrons. By applying the clean room techniques of the semiconductor industry during the assembly and preparation of the cavities one can almost achieve the physical limit of the superconducting material.

In the FLASH linac the cavities are operated at an accelerating field of 23 MV/m. The same value is foreseen for the European X-ray FEL to reach wavelengths below 0.1 nm. Although the quality factor exceeds the excellent value of 10^{10} the RF power dissipation in the cavities would be in the order of 100 kilowatts for the whole linac of the XFEL if the cavities were operated in continuous mode. This is far too large for the liquid helium plant because a heat deposition of 100 kW at $T = 2$ K requires a primary electric power of almost 100 MW at the refrigerator. The necessary reduction of the cryogenic load is the only (and unfortunate) reason to operate the cavities in pulsed mode with a duty cycle of about one per cent. Continuous operation becomes a realistic option if the accelerating field and the electron beam energy are decreased by a factor of 2.5.

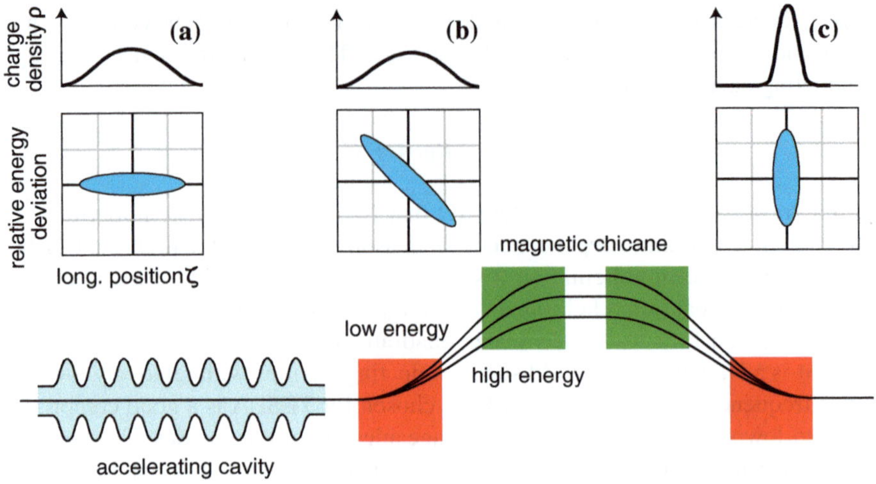

Fig. 8.6 Principle of longitudinal compression of a relativistic electron bunch. The bottom row shows an accelerating cavity and the four dipole magnets of the magnetic chicane. The top figures show the longitudinal bunch charge distribution and the correlation between the longitudinal position ζ of an electron inside the bunch and its relative energy deviation, at three locations: **a** In front of the cavity, **b** behind the cavity, **c** behind the magnetic chicane. In the RF cavity the trailing electrons receive a larger energy gain than the leading ones. In the magnetic chicane the electrons at the tail of the bunch move on a shorter orbit than those at the head and catch up with them. Figure adapted from R. Ischebeck [13].

8.6 Bunch Compression

8.6.1 Principle of Longitudinal Compression

We have seen above that the high peak currents of several 1000 A which are needed in linac-based EUV and X-ray free electron lasers cannot be produced directly in the electron gun because huge space charge forces would disrupt the beam within a short distance. Therefore bunches with a modest peak current are created in the RF photo-injector, accelerated to higher energy and then compressed in length by two orders of magnitude. The electrons in the linac have speeds very close to c, and the velocity differences are too small that a trailing electron would have a chance to catch up with a leading electron if the particles move on a straight line. This possibility is opened if the particles travel through a chicane made of bending magnets.

Longitudinal bunch compression is achieved in two steps: first an energy slope is imprinted on the bunch by off-crest acceleration in the first acceleration module, the particles at the tail of the bunch receiving a larger energy gain than those at the head. Afterwards the particles are passed through a magnetic chicane where the trailing electrons of larger energy travel a shorter distance than the leading ones of smaller energy and are thus enabled to catch up with them. The principle of longitudinal bunch compression is illustrated in Fig. 8.6. Note that the area of the phase-space ellipses shown in Fig. 8.6 is an invariant according to the Liouville Theorem. A length reduction of the bunch is necessarily coupled with an increased energy spread.

8.6.2 Practical Realization of Bunch Compression

Generation of an energy slope

The energy slope needed for magnetic bunch compression is generated by choosing an off-crest RF phase in the accelerating cavities preceding the magnetic chicane. The bunches are accelerated on the falling slope of the RF voltage. Ideally the induced energy variation along the bunch axis should be linear,[2] but this is not what happens in practice because the incoming bunches are too long. Their length amounts to about one per cent of the RF wavelength of 231 mm, and the cosine-like time dependence of the accelerating voltage leads to a pronounced nonlinearity which is clearly visible in the left part of Fig. 8.7. During this measurement the bunch compressor magnets were switched off, so the bunches traveled straight through the linac, retaining their full length of several millimeters. The energy $W(\zeta)$ of the electrons was measured as a function of their longitudinal position ζ inside the bunch using the transversely deflecting microwave structure TDS (see Sect. 8.8.1) in combination with a bending magnet.

A linearization of the accelerating voltage can be achieved by superimposing the 1.3 GHz accelerating field with its third harmonic. A superconducting 3.9 GHz cavity module was developed and built at Fermilab and installed in FLASH in 2010 [14]. With a proper choice of RF amplitude and phase in the third-harmonic cavity module, the so-called "RF curvature" can be removed and the relative energy deviation $\delta W(\zeta)/\overline{W} = (W(\zeta) - \overline{W})/\overline{W}$ becomes an almost linear function of ζ, as shown in the right part of Fig. 8.7. This is a prerequisite for ideal bunch compression.

Two-stage bunch compression

The longitudinal bunch compression in the FLASH linac is accomplished in two steps as shown in Fig. 8.1. The first compression takes place in the magnetic chicane BC1 at an energy of 150 MeV and raises the peak current to several 100 A, the second compression is done in BC2 at an energy 450 MeV and leads to a peak current in the kA range.

A single-stage compression might appear simpler but has severe disadvantages. To understand this we consider two cases: (1) The final peak current of several kA is already produced in BC1. The internal repulsive force (8.4) in a kA-beam is so big at 150 MeV that it causes an intolerable beam blowup and emittance growth before the bunch reaches the next acceleration module. (2) A single-stage compression in BC2 at 450 MeV appears much more favorable in this respect, the repulsive force being a factor of nine smaller. However, the disadvantage of the larger beam energy is that a correspondingly larger energy slope is needed to compress the bunch to

[2] This applies if one neglects a small second-order correction to the beam transfer matrix of the chicane.

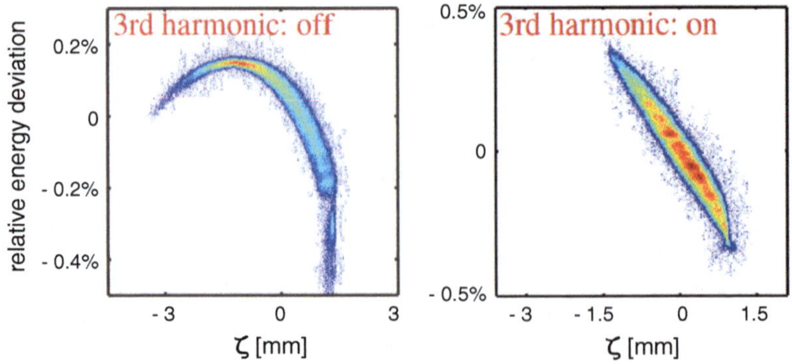

Fig. 8.7 *Left*: Measured relative energy deviation $\delta W(\zeta)/\overline{W}$ as a function of the internal bunch coordinate ζ for a 4 mm long bunch having passed the first acceleration module at an off-crest phase of 6°. $\overline{W}$ is the mean value of the electron energy. The bunch head is to the right. *Right*: Measured relative electron energy deviation as a function of ζ after linearization of the accelerating voltage using the third-harmonic cavity.

its desired short length. The consequence is an excessive beam energy spread after compression, and this energy spread, as we know from Sect. 6.1, has a detrimental effect on the FEL gain. In the two-stage compression scheme, a large fraction of the energy chirp is applied at the lower energy and contributes less to the final relative energy spread σ_η of the beam.

In a bunch compression system without a third-harmonic cavity, the two-stage compression has the additional advantage that the negative impact of the "RF curvature" is significantly smaller than in the single-stage compression.

8.6.3 Simulation of Bunch Compression

Before 2010 the third-harmonic cavity was not available at FLASH. Bunch compression was a rather critical issue and many numerical simulations were carried out to optimize the process. Owing to the cosine shape of the RF wave, the compressed bunches do not possess the desired shape of a narrow peak but consist of a very short leading spike and a long tail. The leading spike contains only 10 – 20 % of the total bunch charge and reaches the peak current of more than 1000 A needed in the high-gain FEL process. Most of the charge is spread out in the tail, where the local current density is too small to expect any significant FEL gain. A simulation of the bunch compression in an early stage of the EUV FEL in Hamburg is shown in Fig. 8.8.

With the help of the third-harmonic cavity it is possible to squeeze a major fraction of the bunch charge into a high-current pulse whose length can be varied between 30 and 400 fs (FWHM). To realize ultrashort bunches with a length of 10 fs or less, a

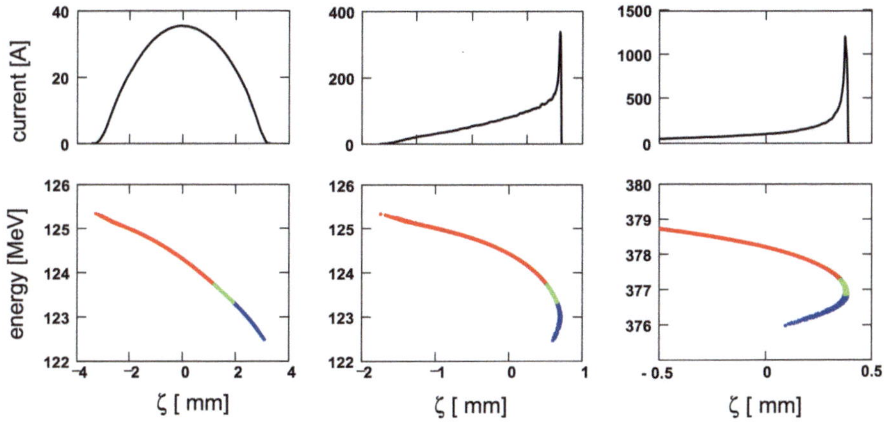

Fig. 8.8 Numerical simulation of two-stage bunch compression without third-harmonic cavity. The bunch charge is 0.5 nC. The figures in the *top row* show the dependence of electron beam current on the internal bunch coordinate ζ. In the *bottom row* the dependence of electron energy on ζ is displayed. *Left*: Behind the first acceleration module. The energy varies from 122.3 MeV at the head of the bunch to 125.3 MeV at the tail. The curvature in the energy chirp is visible. *Middle*: Behind the first magnetic bunch compressor BC1. The lower energy electrons have collected themselves in a narrow peak at the bunch head, the higher energy particles form a long tail. The peak current has grown from 35 to 350 A. *Right*: Behind the second bunch compressor BC2. The mean beam energy is 377 MeV and the peak current has grown to 1250 A. Note that the electrons which were at the front in the left picture are moved backwards in the second bunch compressor due to "over-compression" (*blue color in the plots*).

low charge in the 10–20 pC range is needed, and it is advisable to reduce the length of UV laser pulse at the electron gun, thereby relaxing the requirements on bunch compression.

8.6.4 Collective Effects in the Bunch Compressors

While the principle of magnetic bunch compression is easy to understand—it resembles the compression of chirped laser pulses—there are subtle details which are special for relativistic electrons. The main problem is that short bunches moving on the curved trajectory through the magnetic chicane emit coherent synchrotron radiation (CSR) of high intensity. Radiation emitted at the bunch tail may take a shortcut through the vacuum chamber of the magnetic chicane and interact with the electrons at the bunch head, changing their energy. Also space charge forces have a considerable influence on the shape of the bunches. These interactions and self-forces are not present in photon beams. The strong modifications of bunch shape and internal energy distribution, caused by these collective effects, are demonstrated in Fig. 8.9.

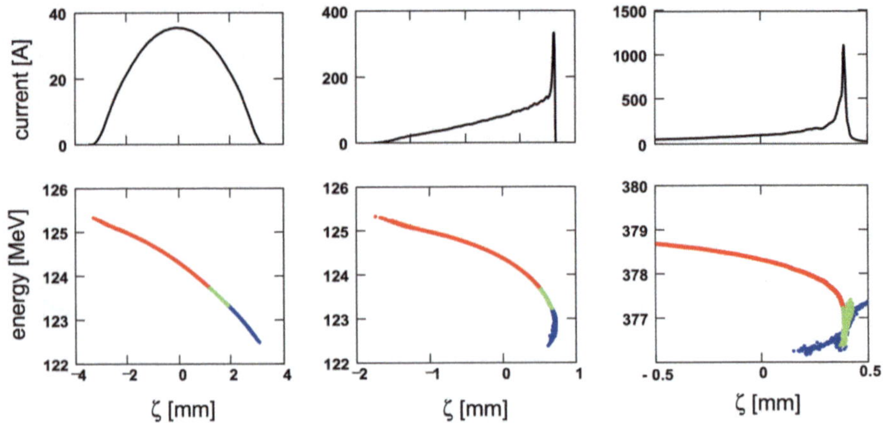

Fig. 8.9 Influence of coherent synchrotron radiation (CSR) and space charge forces on the bunch compression process depicted in Fig. 8.8. The *colors* indicate the initial positions in the bunch. *Blue* head of bunch, *green* middle section, *red* tail.

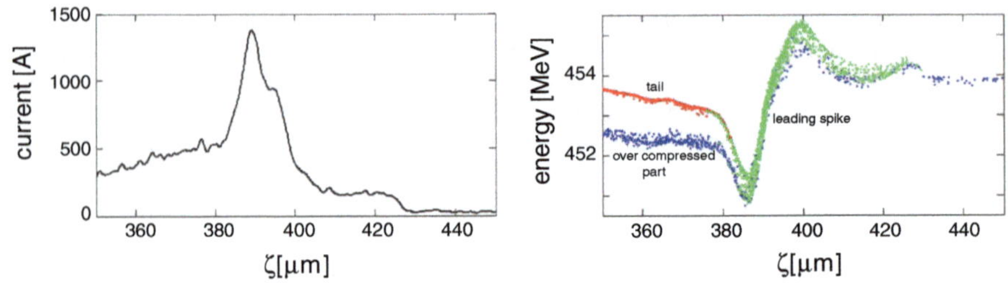

Fig. 8.10 Current profile (*left*) and energy distribution (*right*) in the bunch head region for a 0.5 nC bunch entering the undulator. The mean beam energy is 454 MeV in this simulation.

The bunch shape behind the first bunch compressor is not much affected by the collective effects, but behind the second bunch compressor we observe severe distortions in the shape and energy distribution of the bunch.

The magnetic deflection in front of the undulator (see Fig. 8.1) has an additional influence on the bunch shape. Figure 8.10 shows the computed current profile and energy distribution in the bunch head region. A strong energy variation is observed within the leading peak which results in a broadening of the FEL spectral distribution and a reduction of the FEL gain. The influence of collective effects on the beam properties has been thoroughly investigated in so-called start-to-end simulations [15].

8.7 Wake Field Effects

A fast ship moving down a narrow canal produces a wake that is reflected by the shore and acts back on the ship itself or on other ships in the canal. Similar effects happen when intense relativistic particle bunches travel in the vacuum chamber or the cavities of an accelerator. The charged particles generate electromagnetic fields that are modified by the metallic boundaries. These so-called wake fields may act back on the bunch itself or on trailing bunches and perturb the motion of the particles and change their energy distribution. Wake fields are thoroughly treated in the book *Physics of Collective Beam Instabilities in High Energy Accelerators* by A.W. Chao [16].

We consider first the simplest case of an ultra-relativistic bunch traveling on the axis of a smooth cylindrical beam pipe of circular cross section. The walls are assumed to have perfect conductivity. Since for $\gamma \to \infty$ the electric field lines emerging from the bunch are exactly perpendicular to the direction of motion, these field lines enter the metallic wall at right angles and fulfill thus the boundary condition at the interface between vacuum and a perfect conductor. This means that the field pattern inside the beam pipe is not changed by the presence of the metallic wall. In other words: the metallic boundary does not "exist" for the particle bunch, and wake field effects are absent in this case.[3]

The situation changes if one of the above assumptions is not fulfilled. We consider two important cases

- a finite conductivity of the beam pipe leads to resistive-wall wake fields,
- a variation of the cross section results in geometric wakes.

Resistive-wall wake fields

Resistive-wall wake fields play a role in the 30 m long beam pipe in the undulator system which has a radius of 5 mm and is made from aluminum. The computed longitudinal wake field for a compressed bunch in FLASH is plotted in Fig. 8.11. The wake field is negative in the bunch head but becomes slightly positive in the tail. This means that the electrons in the bunch head lose energy while those in the tail gain energy. The energy spread in the bunch head region, induced by resistive wall wake fields, is estimated as $\sigma_\eta^{\text{wake}} < 0.1\,\rho_{\text{FEL}}$. It is thus uncritical for FLASH but may be a serious concern in an X-ray FEL where the undulator beam pipe is much longer and the FEL parameter significantly smaller than at FLASH. In this case also the beam energy loss caused by resistive-wall wake fields plays a role, see Sect. 9.2.4.

---

[3] In a beam pipe of arbitrary cross section, which is kept invariant along the axis, the field pattern will be more complicated than in a round pipe, but wake field effects are still absent if the resistance of the wall vanishes and the particles are ultra-relativistic.

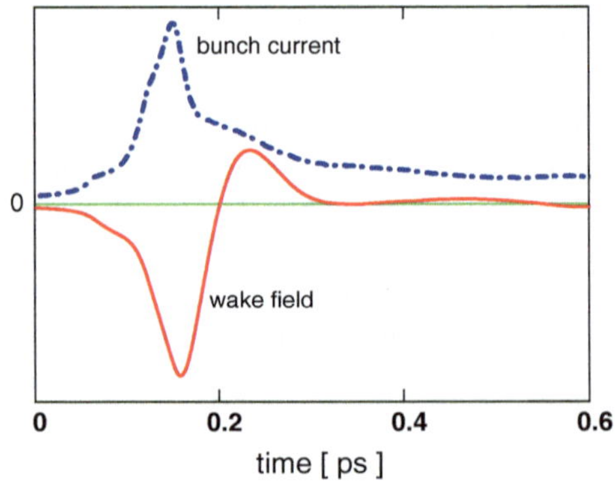

Fig. 8.11 Resistive-wall wake field effect in the aluminum beam pipe of the undulator. The bunch current as a function of time is shown as a *dash-dotted blue curve*. The longitudinal electric wake field is shown as a *continuous red curve*. For a bunch charge of 0.5 nC the maximum bunch current is about 1200 A. The peak value of the wake field amounts to −30 kV/m. The bunch head is at the left side.

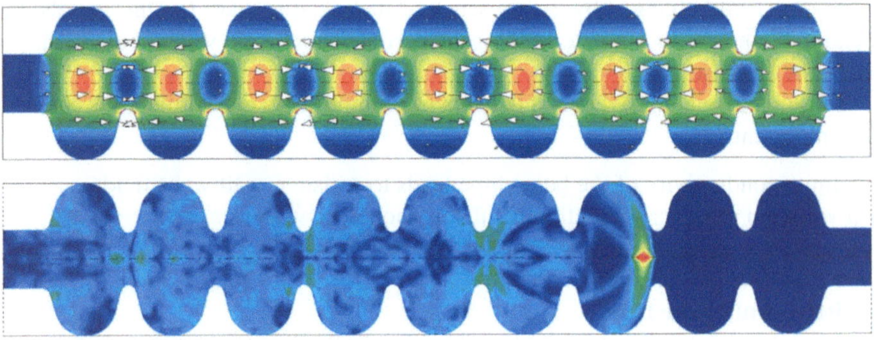

Fig. 8.12 *Top*: The computed field pattern in a nine-cell cavity which is excited to the fundamental TM_{010} mode by a klystron operating at 1.3 GHz. *Bottom*: The wake field pattern generated by a short relativistic electron bunch which is just leaving the seventh cell.

Geometric wakes

Geometric wakes in the nine-cell superconducting cavities, which are excited by the short electron bunches, are shown in the bottom part of Fig. 8.12. Many of the induced higher-order eigenmodes of the cavity are coupled out by a specially designed antenna, and their energy is absorbed in a dump resistor at room temperature. Very high-frequency modes are not trapped and may leave the cavity and travel along the beam pipe. Damping of the traveling modes is a challenging task in a superconducting linear accelerator.

8.8 Longitudinal Electron Beam Diagnostics

The requirements on electron beam quality are very demanding and in some respects at the limit of present-day technology. High-resolution diagnostic instruments are essential for a detailed understanding of the physical principles of emittance preservation, bunch compression, and lasing in the SASE or seeding modes. Moreover, they are an indispensable prerequisite for providing the input signals of the feedback systems for beam energy stabilization and bunch compression.

We restrict ourselves here to a description of three techniques permitting single-shot direct visualization of longitudinal electron bunch profiles with very high resolution: the transversely deflecting microwave structure (TDS), electro-optic (EO) detection systems, and a single-shot infrared spectrometer with 120 wavelength channels.

8.8.1 Transversely Deflecting Microwave Structure

In the TDS the temporal profile of the electron bunch is transferred to a spatial profile on a view screen by a rapidly varying electromagnetic field, analogous to the sawtooth voltage in conventional oscilloscope tubes but providing a thousandfold better time resolution [17, 18]. The TDS used at FLASH was built by SLAC. It is a 3.6 m long traveling-wave structure operating at 2.856 GHz in which a combination of electric and magnetic fields exerts a transverse force for the electron bunches. The bunches pass the TDS near zero crossing of the RF field (phase zero) and receive no net deflection but are streaked in the transverse direction. A single bunch out of a train can be streaked. With a fast kicker magnet, this bunch is deflected towards an observation screen and imaged by a digital camera. The principle of the TDS is explained in Fig. 8.13. The time resolution of the TDS installed at FLASH may be as good as 10 fs (rms), depending on the beam optics chosen.

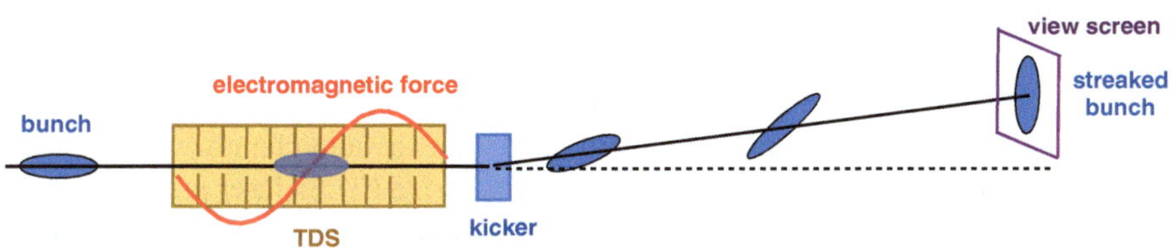

Fig. 8.13 Principle of longitudinal charge density measurement using a transversely deflecting microwave structure. For optimum resolution the RF phase is chosen such that the bunch center coincides with the zero-crossing of the RF wave. This condition holds along the entire axis of the traveling-wave structure since electron bunch and RF wave move synchronously with a speed very close to c.

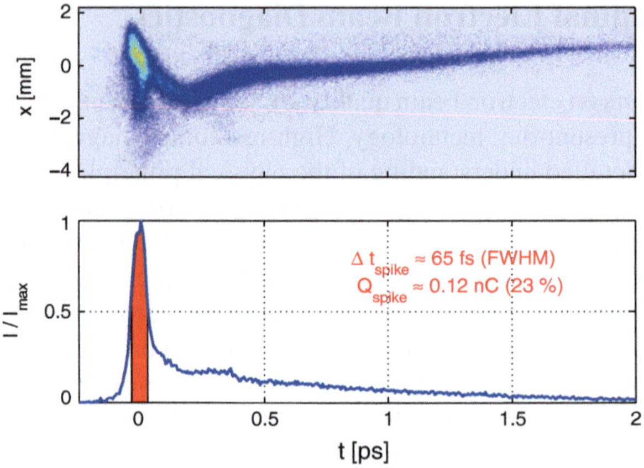

Fig. 8.14 *Top*: Two-dimensional image of a single electron bunch whose time profile is translated into a spatial coordinate on an observation screen. The bunch head is at the left side. *Bottom*: Current as a function of time. The maximum current is $I_{max} = 1.8\,\text{kA}$ in this measurement [19].

An image of a streaked electron bunch is shown in Fig. 8.14. The figure shows also the computed temporal current profile. One observes a sharp peak at the head with a full width at half maximum of 65 fs and a long tail towards later times. The sharp peak contains about 20 % of the bunch charge, and only here is the local charge density large enough to obtain a high gain in the FEL process.

8.8.2 Electro-Optic Detectors

The electro-optic (EO) effect offers the possibility to measure the longitudinal charge distribution in the electron bunches with a resolution of 50 fs. The principle is as follows: the electric field of the relativistic bunch induces an optical birefringence in a crystal such as gallium-phosphide (GaP), which is then probed with femtosecond titanium-sapphire (Ti:Sa) laser pulses. Several EO experiments have been performed in the FLASH linac in the straight section between the last acceleration module and the undulator. The EO crystal is mounted inside the vacuum chamber at a distance of 4 – 5 mm to the electron beam. The linearly polarized Ti:Sa laser pulse enters the chamber at a small angle, crosses the EO crystal and is then coupled out through a quartz window. In the birefringent GaP crystal the laser pulse acquires an elliptical polarization which is transformed into an intensity modulation using a quarter wave plate, a half wave plate and a crossed polarization filter. The Ti:Sa laser produces a periodic pulse train with a repetition rate of 81 MHz. This frequency is adjusted to exactly 1/16 of the 1300 MHz radio-frequency of the linac and then locked to the RF in a phase-locked loop circuit.

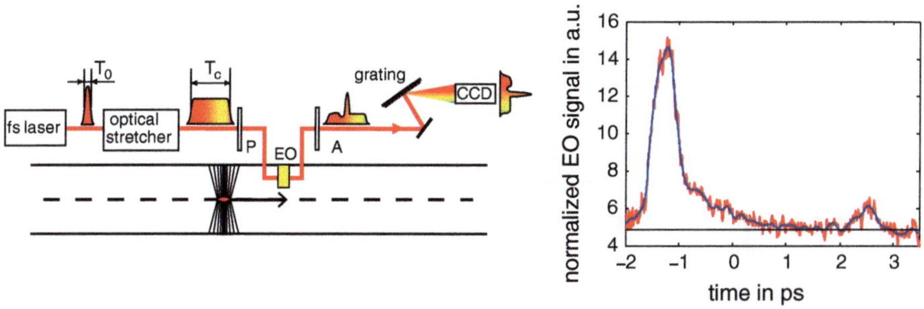

Fig. 8.15 *Left*: Schematic of the electro-optic spectral decoding (EOSD) setup for measuring the time profile of single electron bunches. At the top the chirped laser pulse of 1–2 ps length is shown, at the bottom the radial electric field of the electron bunch. *Right*: Time profile of a compressed electron bunch as measured with the EOSD method. Figure adapted from Ref. [20].

In the electro-optic sampling method, the narrow laser pulses are moved in small steps across the wider electron bunches. Thereby the average time profile of many bunches is obtained. Due to a relative time jitter of 40 fs between the laser pulses and the electron bunches, this method is not adequate for the analysis of sub-100-fs bunches. Several single-shot techniques have been applied permitting the analysis of individual electron bunches, based on spectral, temporal or spatial decoding. The simplest one is the electro-optic spectral decoding (EOSD) method. For this purpose the laser pulse is passed through a dispersive material and stretched (*chirped*) to a length of several ps, longer than the electron bunch. In the chirped pulse the long wavelengths are at the head of the pulse and the short ones at the tail. Note that a 15 fs Ti:Sa laser pulse has a central wavelength of 800 nm and a bandwidth of ±30 nm. In the GaP crystal the temporal structure of the electron bunch is imprinted onto the spectral components of the laser pulse. With a diffraction grating and a gated CCD camera the time information can be recovered. The principle is explained in Fig. 8.15 together with a typical measurement of a single electron bunch [20, 21]. The resolution limit of the spectral decoding method is about 100 fs (rms).

A factor of two better resolution is attainable with the technically more involved electro-optic temporal decoding (EOTD) method. Here an amplified femtosecond Ti:Sa laser pulse is split into two pulses. One of these remains narrow while the other one is stretched to a length of several ps and passes through the EO crystal. The two beams are crossed at finite angle in a nonlinear BBO crystal where they generate second-harmonic (UV) radiation. The time profile of the electron bunch is translated into a spatial variation of the ultraviolet light intensity emerging from the BBO crystal. The ultimate resolution of the EOTD technique of 50 fs (rms) was reached at FLASH in an experiment carried out in cooperation between scientists from the infrared free-electron laser FELIX (The Netherlands), the University of Hamburg and DESY [22]. The experimental setup and a reconstructed electron bunch profile are shown in Fig. 8.16.

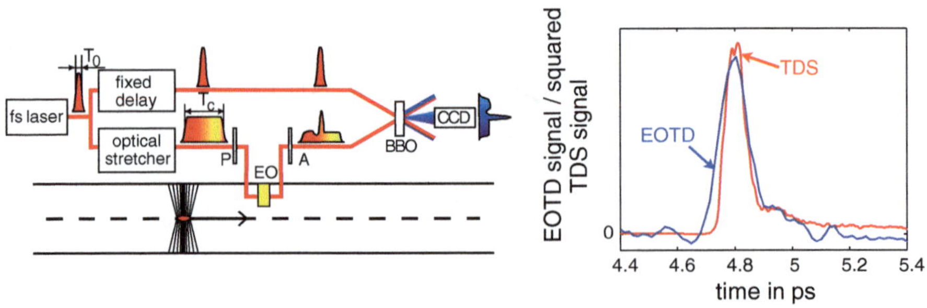

Fig. 8.16 *Left*: Principle of electro-optic temporal decoding (EOTD). *Right*: Comparison of two high-resolution diagnostic methods for single compressed electron bunches. *Red curve* : transversely deflecting structure (TDS), *blue curve*: electro-optic temporal decoding (EOTD). Adapted from Ref. [20].

The EO experiments are inferior to the TDS in terms of time resolution but have the considerable advantage of being "non-destructive": the same bunch which has been analyzed with the EO system can be used to generate FEL radiation downstream. In contrast to this a bunch that has been streaked by the TDS suffers a considerable emittance growth. Moreover, the EO signals can be utilized as arrival time signals of the FEL pulses in *pump-and-probe* experiments [23], but the time resolution offered by the bunch arrival time monitor (Sect. 8.10) is superior.

8.8.3 Coherent Transition Radiation Spectroscopy

Coherent transition radiation (CTR) in the far-infrared regime has a long tradition as a tool for the longitudinal diagnostics of short electron bunches. An interesting application was the reconstruction of a non-symmetric electron bunch shape by autocorrelating the CTR pulse by means of a Martin-Puplett interferometer and computing the phase of the bunch form factor with the help of the Kramers-Kronig dispersion relation [24].

In the Martin-Puplett interferometer the optical delay between the two arms is varied in small steps by moving a mirror. Each bunch makes just one entry in the autocorrelation plot, and many successive bunches are needed to obtain an average longitudinal shape. To overcome this limitation, a novel multichannel infrared and THz spectrometer with fast readout was jointly developed at DESY and the University of Hamburg [25, 26]. The aim was to permit CTR spectroscopy on single electron bunches.

Multichannel infrared spectrometer

In the FLASH linac, transition radiation is produced on a metalized screen inside the ultrahigh vacuum beam pipe of the linac, coupled out through a diamond win-

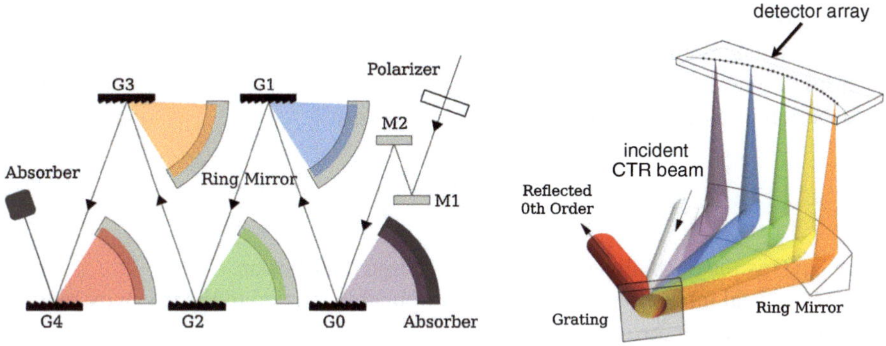

Fig. 8.17 *Left*: Principle of the multi-channel infrared spectrometer equipped with four consecutive blazed reflection gratings, G1–G4, each covering about a factor of two in wavelength. *Right*: Layout of one spectrometer stage, comprising the grating, a ring mirror and an array of 30 pyroelectric detectors. The light dispersion and focusing have been computed with a ray tracing code. For clarity only 5 of the 30 wavelength channels are shown.

dow with high transmission from 200 GHz up to optical frequencies, and transported to a laboratory through a CTR beamline [27] comprising six focusing mirrors that are mounted in an evacuated tube. The infrared spectrometer is mounted in a vacuum chamber directly attached to the CTR beam line, and it is equipped with four consecutive dispersion gratings and a total of 120 parallel readout channels. The spectrometer can be operated either in the long-wavelength mode (45–430 μm) or in the short-wavelength mode (5–44 μm). The layout is shown in Fig. 8.17. Pyroelectric detectors with a short time constant, fast amplifiers and ADCs with parallel readout permit the broadband spectroscopy of coherent radiation from single electron bunches.

The operational principle of the spectrometer is described for the far-infrared mode. The incident radiation is passed through a polarization filter to select the polarization component perpendicular to the grooves of the gratings, and is then directed towards grating G0 which acts as a low-pass filter: high-frequency radiation (wavelength $\lambda < 44\,\mu$m) is dispersed and guided to an absorber, while for low frequencies ($\lambda > 45\,\mu$m) the grating G0 acts as a plane mirror and directs the radiation to the first grating stage G1 of the spectrometer. At grating G1, radiation in the range from 45.3 to 77.4 μm is dispersed and focused onto a multi-channel detector array, while radiation with $\lambda > 77.6\,\mu$m is specularly reflected and sent to G2. The subsequent gratings work similarly and disperse the wavelength intervals 77.0–131.5 μm (G2), 140.0–239.1 μm (G3), and 256.7–434.5 μm (G4). For each grating, the first-order diffracted radiation is focused by a ring mirror of parabolic shape onto an array of 30 pyroelectric detectors which are arranged on a circular arc. This is shown schematically in Fig. 8.17. Technical details and the measured performance are reported in Ref. [26].

Method of bunch shape reconstruction
The internal longitudinal coordinate ζ in the bunch can be translated into a time coordinate via the relation $\zeta = -\beta ct \approx -ct$. This allows us to write the charge distribution of a bunch with N electrons in the form

$$\rho(t) = -N\, e\, \rho_n(t) \quad \text{with} \quad \int_{-\infty}^{\infty} \rho_n(t) dt = 1\,. \tag{8.8}$$

Here $\rho_n(t)$ is the normalized longitudinal charge distribution.

The spectral intensity radiated by such a bunch is related to the intensity radiated by a single electron through the formula

$$I_N(\omega) = I_1(\omega) \left(N + N(N-1)\, |\tilde{F}(\omega)|^2 \right), \tag{8.9}$$

where $\tilde{F}(\omega)$ is the longitudinal form factor of the bunch, the Fourier transform of the normalized longitudinal charge distribution:

$$\tilde{F}(\omega) = \int_{-\infty}^{\infty} \rho_n(t) e^{i\,\omega\, t} dt = \int_{-\infty}^{\infty} \rho_n(t)(\cos\omega\, t + i\, \sin\omega\, t)\,. \tag{8.10}$$

In writing down this equation we have made use of the fact that transition radiation from highly relativistic electrons is confined to small angles which implies that the influence of the transverse beam size is strongly suppressed but not entirely negligible at small wavelengths. The longitudinal form factor has to be corrected for the finite transverse beam size, see [26, 28].

At sufficiently low frequencies, namely when the wavelength of the radiation is long compared to the bunch length, the form factor is a real number and approaches unity in the limit $\omega \to 0$. Then all electrons radiate coherently which means that there is constructive interference among their radiation fields. Spectral measurements in this range yield practically no information on the internal charge distribution in the bunch. To gain such information, measurements at wavelengths shorter than the bunch length have to be carried out. In that case the interference pattern is partially constructive, partially destructive and the form factor becomes a complex number

$$\tilde{F}(\omega) = F(\omega)\, e^{i\,\phi(\omega)}\,, \tag{8.11}$$

whose magnitude $F(\omega) = |\tilde{F}(\omega)|$ is generally less than 1. If both $F(\omega)$ and $\phi(\omega)$ were measurable, a unique reconstruction of the charge distribution $\rho(t)$ could be achieved by the inverse Fourier transformation. Unfortunately only the spectral intensity $I_N(\omega)$ is measurable in spectroscopic experiments at accelerators. Hence the modulus $F(\omega)$ of the longitudinal form factor can be determined, while the phase $\phi(\omega)$ remains unknown.

A similar problem arises for the optical reflection properties of solids. In this case it is possible to compute the phase of the complex reflectivity amplitude from the measured reflectivity by means of the Kramers-Kronig relation which connects the real and imaginary parts of a complex analytic function. It can be shown that the reflectivity amplitude is an analytic function of the complex frequency $\tilde{\omega} = \omega + i\,\kappa$. A rigorous mathematical treatment is presented in the book *Optical Properties of Solids* by F. Wooten [29]. The method has been adopted for the phase reconstruction of the complex bunch form factor, see the article by Lai and Sievers [30] and the references quoted therein. The phase $\phi(\omega_0)$ can be computed from the real function $F(\omega)$ by means of the following integral

$$\phi(\omega_0) = \frac{2\omega_0}{\pi} \int_0^\infty \frac{\ln F(\omega) - \ln F(\omega_0)}{\omega_0^2 - \omega^2}\, d\omega\,. \tag{8.12}$$

It must be noted though that this phase reconstruction suffers from some mathematical problems[4] and has to be considered an empirical method. For example, let the bunch shape be given by the sum of two Gaussians of different width and central position. The reconstruction turns out perfect if the narrow peak comes first, but the sequence of the reconstructed peaks is interchanged when the wide peak comes first. This interchange is easy to understand. The charge density $\rho(t)$ is a real function, so from Eq. (8.10) follows immediately that the real part of the form factor is an even function of ω while the imaginary part is an odd function. The absolute magnitude of the form factor is also an even function of ω, the phase is an odd function:

$$F(-\omega) = +F(+\omega)\,, \quad \phi(-\omega) = -\phi(+\omega)\,. \tag{8.13}$$

An interchange of the two Gaussians corresponds to a time inversion and is equivalent to a sign reversal of ω. This leaves the absolute magnitude of the form factor invariant and has thus no effect on the phase (8.12).

A more serious problem arises for a bunch shape consisting of three Gaussians with increasing width. Here an interchange of the first two peaks leads to a reconstructed shape in which not only the time order of the first two peaks is inverted but in addition their shape is modified. Similar difficulties were reported in [30].

Such ambiguities cannot be avoided. In fact, in Ref. [31] it has been proved that a unique shape reconstruction is not possible. A simple counterexample demonstrates this in a convincing way. The following two functions are considered

$$f_1(t) = f_2(t) = 0 \ \text{ for } \ t < 0\,,$$
$$f_1(t) = e^{-\beta t}\,, \ f_2(t) = e^{-\beta t}\left(1 + \frac{4\beta^2(1-\cos(\alpha t)}{\alpha^2} - \frac{4\beta\sin(\alpha t)}{\alpha}\right) \ \text{ for } \ t \geq 0$$

with real parameters $\alpha, \beta > 0$. The first function has an infinitely steep rise at $t = 0$, followed by an exponential decay. The second function has the same steep rise but

[4] The bunch form factor may have zeroes in the upper half of the complex $\tilde{\omega}$ plane which lead to singularities of $\ln[F(\tilde{\omega})]$ and contribute additional terms to the right-hand side of Eq. (8.12). These contributions are not experimentally accessible. For a further discussion see [30].

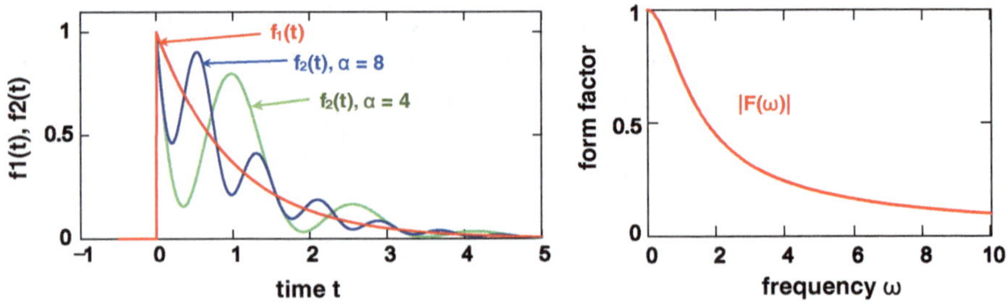

Fig. 8.18 *Left*: The functions $f_1(t)$ and $f_2(t)$ for $\beta = 1$ and $\alpha = 4$ resp. $\alpha = 8$. *Right*: The modulus of the form factors, $|\tilde{F}_1(\omega)| = |\tilde{F}_2(\omega)|$. We thank B. Schmidt for numerous discussions on the phase reconstruction problem.

the decay is superimposed with an oscillatory pattern, see Fig. 8.18. The complex form factors $\tilde{F}_1(\omega)$ and $\tilde{F}_2(\omega)$ differ, but surprisingly their absolute magnitudes are identical

$$|\tilde{F}_1(\omega)| = |\tilde{F}_2(\omega)| = (\omega^2 + \beta^2)^{-1/2} .$$

The Kramers-Kronig method reproduces the function $f_1(t)$ very accurately, but any attempt to reconstruct the oscillatory function $f_2(t)$ fails, no matter what the value of α is. One obtains always the non-oscillatory function $f_1(t)$. From this and other examples one can conclude that the Kramers-Kronig method tends to generate a time profile possessing the minimum amount of structure that is compatible with the measured CTR spectrum.

Notwithstanding these fundamental limitations, the phase reconstruction method yields valuable supplementary information, in particular if the overall features of the longitudinal bunch shape are known from a TDS measurement.[5] Then the spectroscopic data at high frequency can be utilized for an improvement of the time resolution. It is indeed the combination of these two complementary methods which improves the reliability and faithfulness of the bunch shape determination.

Reconstructed electron bunch shapes

In order to demonstrate the capabilities of the CTR spectrometer, spectral measurements on short bunches are presented in the left part of Fig. 8.19. Shown is the magnitude of the form factor, derived from the combined infrared spectra of two measurement series on single bunches: (a) 100 bunches with the spectrometer in the short-wavelength mode and (b) 100 bunches with the spectrometer in the long-wavelength mode. A broad wavelength spectrum is observed, ranging from 9 μm up to 450 μm. In the right part of Fig 8.19, the bunch time profile derived from these data

[5] It should be noted that also the TDS data suffer from ambiguities, the reconstructed shape depends on the streak direction. A possible explanation is that the particles possess a non-vanishing average slope $\langle y'(\zeta)\rangle$ in streak direction which varies along the bunch axis. The resulting shape errors cancel when taking the average of the TDS measurements with positive and negative streak direction.

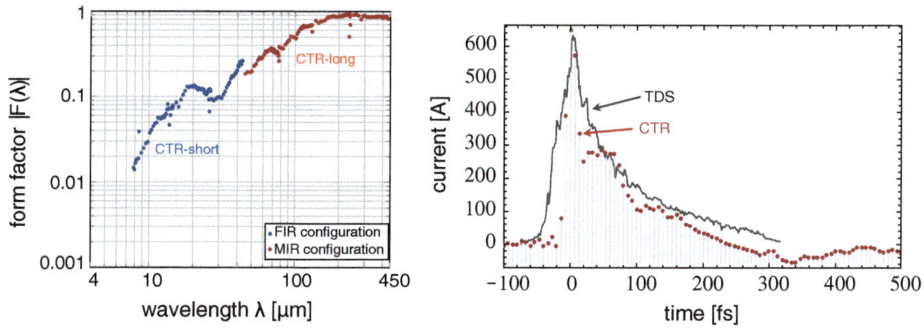

Fig. 8.19 *Left*: Form factor $|F(\lambda)|$, derived from the combined infrared spectra from two measurement series on single bunches: **a** 100 bunches with the spectrometer in the short-wavelength mode (*blue points*) and **b** 100 bunches with the spectrometer in the long-wavelength mode (*red points*). *Right*: Comparison of the bunch time profile derived from the CTR data with a time-domain measurement using the TDS. (Courtesy of B. Schmidt).

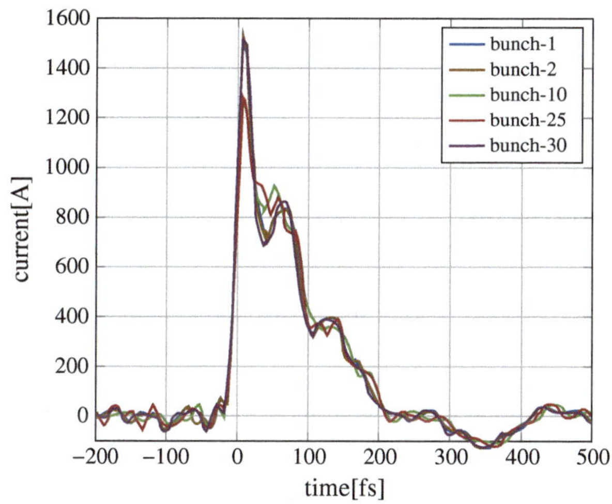

Fig. 8.20 Reconstructed time profiles of 5 selected bunches, combining the single-shot spectra obtained in the short-wavelength mode with the averaged spectrum measured in the long-wavelength mode. (Courtesy of E. Hass and S. Wesch).

is compared with a TDS measurement. There is good overall agreement between the two complementary measurements.

The CTR spectrometer permits a quasi-single-shot determination of the longitudinal bunch shape. For this purpose, the averaged long-wavelength spectrum, depicted in Fig. 8.19, is combined with the single-shot spectra in the short-wavelength mode. The computed time profiles are displayed in Fig. 8.20. They show a sharp leading spike followed by a long tail. The shot-to-shot reproducibility is quite remarkable. The CTR system is thus well suited for online monitoring of the bunch shape during FEL user runs.

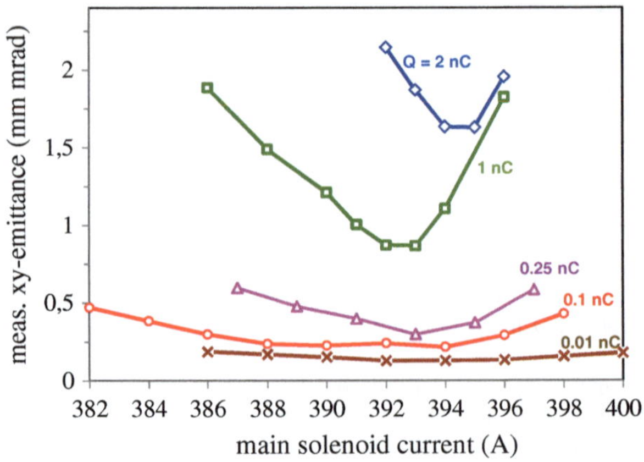

Fig. 8.21 The normalized horizontal and vertical emittance as a function of the main solenoid current in the electron gun (courtesy of M. Krasilnikov). The bunch charge varies between 0.01 and 2 nC. The data were obtained at the DESY-Zeuthen photo-injector test facility for an uncompressed electron beam with an energy of 25 MeV (for details see Ref. [10]).

8.9 Transverse Beam Diagnostics

Emittance measurements

A standard method for determining the emittance is based on measuring the two-dimensional transverse intensity profile of the beam at various positions in the linac. For this purpose view screens are inserted that intercept the beam. At FLASH these are either thin polished silicon discs with an aluminum coating on the front face which generate optical transition radiation, or alternatively thin Ce:YAG crystals which produce scintillation light. Two-dimensional images are recorded with digital cameras. From the known beam-optical transfer matrices between the observations screens, and by varying the quadrupole strengths, the particle distribution in phase space can be tomographically reconstructed and the beta function and the beam emittance can be determined. As an example, Fig. 8.21 shows the normalized transverse emittance as a function of the current in the main solenoid magnet of the electron gun. This figure demonstrates how critically the beam focusing in the electron injector depends on the focal properties of the solenoid coil. Even more critical is the bunch charge. For bunch charges below 0.1 nC the minimum value of the normalized emittance is $\varepsilon_n \leq 0.5\,\mu\text{m}$, but it grows to more than $1\,\mu\text{m}$ when the bunch charge is raised beyond 1 nC. The emittance criterion (6.17) is well satisfied for an uncompressed beam. However the nonlinear phenomena in the bunch compressors lead to an appreciable emittance growth if the bunch charge is high.

Slice emittance

What is really of interest for a high-gain FEL is not the average emittance of the entire bunch but the emittance of the high-current region in the bunch head. Standard emit-

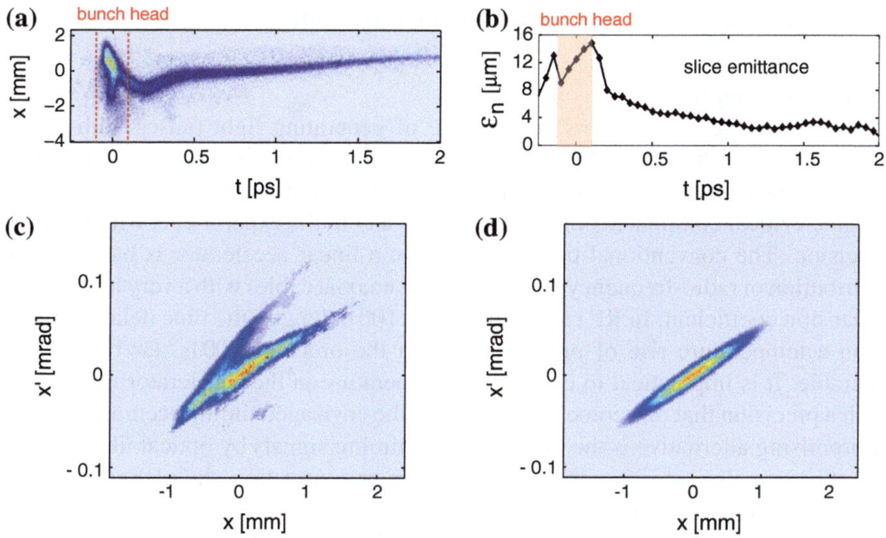

Fig. 8.22 **a** Two-dimensional image (t, x) of a single electron bunch. The bunch head is at the left side. **b** The normalized horizontal emittance as a function of the temporal position in the bunch. **c** Horizontal phase space distribution in the bunch head region. The long wings extending appreciably beyond the central elliptic region are the main source for the large emittance in the bunch head region (shaded area in graph (**b**)). **d** The central elliptic phase space distribution in the bunch head region. The peak current is about 1.4 kA, the normalized emittance is $\varepsilon_n = 4\,\mu$m. See Refs. [19, 32] for details.

tance measurements are incapable of extracting this information. The transversely deflecting microwave structure TDS permits the determination of the emittance in selected time slices of the bunches. For this purpose the quadrupole strengths in the beam optics section between the TDS and the observation screen are varied.

A thorough phase space analysis was carried out for bunches with a time profile as shown in Fig. 8.14. Some results are presented in Fig. 8.22. An unusually large normalized emittance of $\varepsilon_n \approx 10\,\mu$m was derived in the bunch head, a number which is far too large to be compatible with the high FEL gain measured with these machine parameters. The reason is that in this special case "over-compression" was applied (compare Fig. 8.8) leading to a strong deformation of the phase space distribution. The complexity of the data is illustrated in Fig. 8.22c where the reconstructed horizontal phase space distribution is shown for a slice in the bunch head region. Large deviations from the ideal elliptic shape are observed. The long wings extending far beyond the central elliptic region towards large values of x and x' give the main contribution to the large normalized emittance, however these wings have too low a charge density to yield an appreciable FEL gain. If we restrict ourselves to the central elliptic region with high charge density we obtain a normalized emittance of a few μm at an acceptable peak current of about 1.4 kA. This central region is indeed responsible for the observed large FEL gain.

8.10 Laser-Optical Synchronization System

General description
High-gain free-electron lasers are capable of generating light pulses with a time duration of less than 10 fs. In order to fully exploit the vast scientific opportunities that are opened by these ultrashort light pulses, it is highly desirable to synchronize all time-critical components in the accelerator and in the experiments with the same precision. The conventional timing method in a linear accelerator is based on the distribution of radio-frequency signals through coaxial cables with a very low thermal expansion coefficient. In RF cables of several 100 m length, the time delay resulting from a temperature rise of only 0.1 °C is in the order of 100 fs, far bigger than tolerable. It is impractical to control the temperature in the accelerator tunnel with such a precision that one comes even close to the envisaged timing accuracy of 10 fs. A promising alternative is the distribution of timing signals by optical fibers. These are of course also subject to thermal expansion or contraction, their length changes, however, can be compensated, as explained below.

Following a concept proposed at MIT [33], a laser-optical synchronization system has been developed featuring a stability of better than 10 fs. The timing reference is provided by a mode-locked erbium-doped laser operating at the telecommunication wavelength of 1550 nm. The laser produces a periodic train of infrared pulses with a duration of 100 to 200 fs. The timing information is encoded in the highly accurate pulse repetition frequency of 216.66 MHz, one sixth of the 1300 MHz radio-frequency of the accelerating cavities. To ensure the long-term stability of the repetition rate, the laser is phase-locked to the RF master oscillator. The remaining timing jitter is only a few femtoseconds.

The laser pulses are distributed to remotely located end-stations via optical fibers with a length of up to 500 m. The fibers are dispersion-compensated to preserve the shortness of the laser pulses. At the end of the fiber-link, part of the laser power is reflected and sent back through the same fiber. The returning pulse train is superimposed with the incoming laser pulse train in a balanced optical cross-correlator to determine any change in the temporal overlap between the two pulse trains. The cross-correlator generates the input signal for a feedback loop. Variations of the pulse transit time through the optical fiber, caused by thermal expansion resp. contraction or by vibrations of the fiber, are compensated by adjusting the fiber length with the help of a piezo-electric fiber stretcher in combination with an optical delay stage. Transit time changes can be compensated with femtosecond precision.

The periodic laser pulse train is utilized for various applications.

(1) Generation of 1300 MHz RF signals
When the periodic train of narrow IR pulses is detected with a photodiode, the diode output signal contains harmonics at all integer multiples of the repetition frequency of 216.66 MHz. A straightforward way of extracting a 1300 MHz RF signal is to select the sixth harmonic with a bandpass filter. In principle, an RF signal with a stability of better than 10 fs can be generated this way, but attention must be paid to

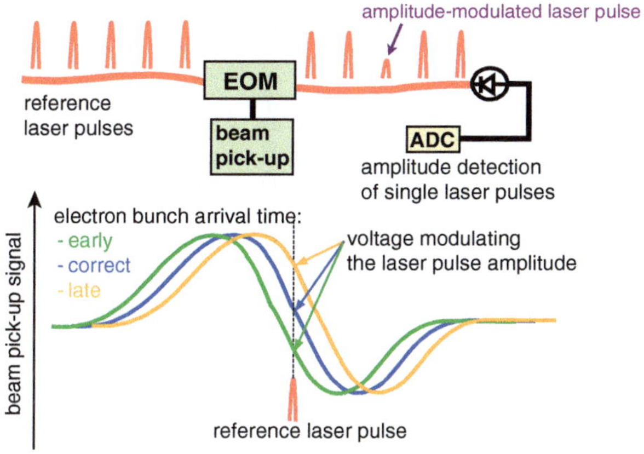

Fig. 8.23 Principle of the bunch arrival time monitor [36]. A periodic train of laser pulses is passed through an electro-optical modulator EOM. If a laser pulse coincides with an electron bunch, it is amplitude-modulated by the beam pick-up signal. A high timing accuracy is obtained by working in the zero-crossing scheme.

choose photodiodes that are insensitive to drifts in temperature and laser power. The resulting systematic time shifts can be avoided employing a Sagnac or Mach-Zehnder interferometer [34].

(2) Synchronization with other lasers
Different types of conventional lasers are in use at the FEL: the UV laser of the photo-injector, the lasers for electro-optic detectors, the laser needed for the seeding operation, and last but not least, lasers in *pump-and-probe* experiments. These mode-locked lasers can be synchronized to the optical pulse train with femtosecond precision by similar cross-correlation methods as used for the length stabilization of the fiber links. A long-term stability of better than 10 fs has been demonstrated [35].

(3) Bunch arrival-time monitor
The arrival-times of single electron bunches at selected locations in the linac are measured with a special device called bunch arrival time monitor (BAM). The signal of a beam pick-up antenna with GHz bandwidth is sampled by the periodic laser pulse train using a broadband electro-optical modulator (EOM). Bunch arrival time deviations are converted into amplitude modulations of the sampling lasers pulses, which are then recorded in a photo-detector. The principle of the monitor is sketched in Fig. 8.23.

Experimental data are shown in Fig. 8.24. In a train of 30 bunches, a systematic variation of the arrival time is visible. The most likely cause is an energy variation in the bunch train which translates into a time variation when the particles travel through the magnetic chicanes of the bunch compressors. There is an excellent correlation between the arrival times at two different BAM positions. From the scattering around

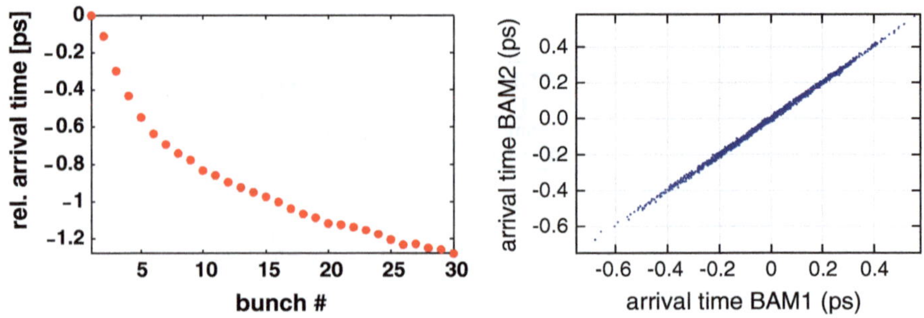

Fig. 8.24 *Left*: Arrival time variation in a train of 30 bunches. *Right*: Correlation between the arrival times at two BAM stations (courtesy of F. Löhl).

the correlation line, the rms time resolution of the bunch arrival-time monitor is estimated to be 6 fs [37].

References

1. L. Lilje et al., Achievement of 35 MV/m in the superconducting nine-cell cavities for TESLA. Nucl. Instrum. Meth. A **524**, 1 (2004)
2. R. Brinkmann, G. Materlik, J. Rossbach, J. Schneider, B.H. Wiik, An X-ray FEL laboratory as part of a linear collider design. Nucl. Instrum. Meth. A **393**, 88 (1997)
3. J. Rossbach, A VUV free electron laser at the TESLA test facility at DESY. Nucl. Instrum. Meth. A **375**, 269 (1996)
4. J. Pflüger, U. Hahn, B. Faatz, M. Tischer, Undulator system for the VUV-FEL at the TESLA test facility phase-2. Nucl. Instrum. Meth. A **507**, 228 (2003)
5. J. Fraser, R. Sheffield, E.R. Gray, A new high-brightness electron injector for free-electron lasers driven by rf linacs. Nucl. Instrum. Meth. A **250**, 71 (1986)
6. I. Will, H.I. Templin, S. Schreiber, W. Sandner, Photoinjector drive laser of the FLASH FEL. Opt. Express **19**, 23770 (2011)
7. B.E. Carlsten, New photoelectric injector design for the Los Alamos National Laboratory XUV FEL accelerator. Nucl. Instrum. Meth. A **285**, 313 (1989)
8. L. Serafini, J.B. Rosenzweig, Envelope analysis of intense relativistic quasilaminar beams in rf photoinjectors: a theory of emittance compensation. Phys. Rev. E **55**, 7565 (1997)
9. K. Flöttmann, ASTRA, A Space charge Tracking Algorithm. http://www.desy.de/~mpyflo (1999)
10. M. Krasilnikov et al., Experimentally minimized beam emittance from an L-band photoinjector. Phys. Rev. ST Accel. Beams **15**, 100701 (2012)
11. B. Aune et al., The superconducting TESLA cavities. Phys. Rev. ST Acc. Beams **3**, 92001 (2000)
12. P. Schmüser, Superconductivity in high energy particle accelerators. Prog. Part. Nucl. Phys. **49**(1), 155 (2002)
13. R. Ischebeck, Transverse coherence of a VUV free electron laser, Ph.D. thesis, University of Hamburg, 2003
14. H. Edwards, C. Behrens, E. Harms, 3.9 GHz cavity module for linear bunch compression at FLASH, in *Proceedings of Linear Accelerator Conference 2010*, Tsukuba, Japan, 2010

15. I. Zagorodnov, M. Dohlus, Semianalytical modeling of multistage bunch compression with collective effects. Phys. Rev. ST Acc. Beams **14**, 014403 (2011)
16. A.W. Chao, *Physics of Collective Beam Instabilities in High Energy Accelerators* (John Wiley, New York, 1993)
17. P. Krejcik et al., Commissioning of the SSPS linac bunch compressor, in *Proceedings of Particle Accelerator Conference PAC 2003*, Portland, Oregon (2003), p. 423
18. R. Akre et al., Bunch length measurements using a transverse RF deflecting structure in the SLAC linac, in *European Particle Accelerator Conference EPAC 2002*, Paris 2002
19. M. Röhrs, Investigations of the phase space distributions of electron bunches at the FLASH-linac using a transverse deflecting structure, Ph.D. thesis, University of Hamburg, April 2008
20. B. Steffen, Electro-optic methods for longitudinal bunch diagnostic at FLASH, Ph.D. thesis, University of Hamburg 2007
21. B. Steffen et al., Electro-optic time profile monitors for femtosecond electron bunches at the soft x-ray free-electron laser FLASH. Phys. Rev. ST Acc. Beams **12**, 032802 (2009)
22. G. Berden et al., Benchmarking of electro-optic monitors for femtosecond electron bunches. Phys. Rev. Lett. **99**, 164801 (2007)
23. A. Azima et al., Time-resolved pump-probe experiments beyond the jitter limitations at FLASH. Appl. Phys. Lett. **94**, 144102 (2009)
24. R. Lai, U. Happek, A.J. Sievers, Measurement of the longitudinal asymmetry of a charged particle bunch from the coherent synchrotron or transition radiation spectrum. Phys. Rev. E **50**, R4294 (1994)
25. H. Delsim-Hashemi et al., Single-shot diagnostics with THz radiation at the free-electron laser FLASH, in *International Free Electron Laser Conference*, Berlin, 2006
26. S. Wesch et al., A multi-channel THz and infrared spectrometer for femtosecond electron bunch diagnostics by single-shot spectroscopy of coherent radiation. Nucl. Instrum. Meth. A **665**, 40 (2011)
27. S. Casalbuoni et al., Ultrabroadband terahertz source and beamline based on coherent transition radiation. Phys. Rev. ST Acc. Beams **12**, 030705 (2009)
28. S. Wesch, Echtzeitbestimmung longitudinaler Elektronenstrahlparameter mittels absoluter Intensitäts- und Spektralmessung einzelner kohärenter THz Strahlungspulse, Ph.D. thesis, University of Hamburg, 2012
29. F. Wooten, *Optical Properties of Solids* (Academic Press, New York, 1972)
30. R. Lai, A.J. Sievers, On using the coherent far IR radiation produced by a charged-particle bunch to determine its shape: I analysis. Nucl. Instrum. Meth. A **397**, 221 (1997)
31. E.J. Akutowicz, On the determination of the phase of a Fourier integral. Trans. Amer. Math. Soc. **83**, 179 (1956)
32. M. Röhrs et al., Time-resolved electron beam phase space tomography at a soft x-ray free-electron laser. Phys. Rev. ST Acc. Beams **12**, 050704 (2009)
33. J. Kim et al., Large-Scale timing distributions and RF synchronization for FEL facilities, in *Proceedings of FEL Conference 2004*, Trieste, Italy (2004), p. 339
34. J. Kim et al., Drift-free femtosecond timing synchronization of remote optical and microwave sources. Nature Photonics **2**, 733 (2008)
35. S. Schulz et al., All-optical synchronization of distributed laser systems at FLASH, in *Proceedings of PAC09*, Vancouver, Canada (2009), p. 4174
36. F. Löhl, Optical Synchronization of a Free-Electron Laser with Femtosecond Precision, Ph.D. thesis University of Hamburg, 2009
37. F. Löhl et al., Electron bunch timing with femtosecond precision in a superconducting free-electron laser. Phys. Rev. Lett. **104**, 144801 (2010)

Chapter 9
X-Ray Free-Electron Lasers: Technical Realization and Experimental Results

The physical and technological challenges of FELs become quite demanding with decreasing wavelength, but in recent years X-ray FELs with wavelengths in the Ångstrøm regime (1 Å = 0.1 nm $= 10^{-10}$ m) have become a reality with the successful commissioning and operation of the "Linac Coherent Light Source" LCLS [1], the world's first FEL providing atomic resolution. In this chapter, we discuss the most important aspects and challenges of X-ray FELs and present some of the excellent experimental results achieved at LCLS and the second facility of this kind, the "Spring-8 Angstrom Compact free-electron LAser" SACLA [2]. Low-gain FEL oscillators in the X-ray regime have been proposed [3] but not yet demonstrated, they are not considered here. Comprehensive information and technical details on X-ray FELs are found in the design reports of LCLS [4] and the European XFEL [5]. Further valuable information is presented in a review article by Huang and Kim [6].

9.1 Photon Beam Brightness

One of the most exciting aspects of an X-ray FEL is its extremely high peak brightness (also called *brilliance*) which is more than eight orders of magnitude higher than in any other X-ray source. The spectral brightness describes the intensity of a radiation source taking into account its spectral purity and opening angle (see [7])

$$B = \frac{\Phi}{4\pi^2 \Sigma_x \Sigma_{\theta x} \Sigma_y \Sigma_{\theta y}}, \tag{9.1}$$

where Φ is the spectral photon flux defined as the number of photons per second and within a given relative spectral bandwidth $\Delta\omega_\ell/\omega_\ell$. The brightness determines how much monochromatic radiation power can be focused onto a tiny spot on the target.

For radiation sources with only partial transverse coherence such as wigglers and undulators at storage rings, the quantities $\Sigma_x, \ldots$ are calculated from the transverse rms sizes and angular divergences of the photon and the electron beam [7]:

P. Schmüser et al., *Free-Electron Lasers in the Ultraviolet and X-Ray Regime*, Springer Tracts in Modern Physics 258, DOI: 10.1007/978-3-319-04081-3_9,

$$\Sigma_x = \sqrt{\sigma_{x,\,ph}^2 + \sigma_{x,\,e}^2}\ , \quad \Sigma_{\theta x} = \sqrt{\sigma_{\theta x,\,ph}^2 + \sigma_{\theta x,\,e}^2} \tag{9.2}$$

and Σ_y, $\Sigma_{\theta y}$ likewise. For a photon beam with full transverse coherence, the transverse size and angular divergence are no longer independent quantities. According to Eq. (10.76) in Sect. 10.3 we get

$$\sigma_x \cdot \sigma_{\theta x} = \frac{\lambda}{4\pi} \tag{9.3}$$

for a photon beam in the fundamental Gaussian (TEM_{00}) mode. The expression in Eq. (9.3) is usually called the *diffraction limit.*

Since all FELs operate in pulsed mode, one has to distinguish between peak brightness, the brightness during the short duration of the photon pulse, and average brightness. For many scientific applications the peak brightness is the figure of merit. The measured peak brightness values of the existing X-ray FELs LCLS and SACLA are depicted in Fig. 9.1. The peak brightness of the EUV and soft X-ray FELs FLASH [8] and FERMI@Elettra [9] is somewhat lower, owing to the larger wavelength, but again many orders of magnitude higher than that of third-generation synchrotron light sources, which are usually storage rings equipped with undulators. Short-wavelength high-gain FELs are often referred to as the fourth-generation of accelerator-based light sources.

In an X-ray FEL equipped with a very long undulator, the fundamental Gaussian mode will eventually dominate, compare Fig. 7.8. If this happens the electron beam properties drop out and one can write

$$\Sigma_x \Sigma_{\theta x} = \Sigma_y \Sigma_{\theta y} = \frac{\lambda_\ell}{4\pi}\ .$$

Hence the brightness of the FEL is simply inversely proportional to the square of the photon wavelength

$$B_{\mathrm{FEL}} = \frac{4\Phi}{\lambda_\ell^2}\ , \tag{9.4}$$

provided the photon flux is independent of wavelength. Indirectly, the electron beam parameters play of course an essential role because stringent upper limits on the transverse size and divergence must be obeyed in order to achieve a high FEL gain and the formation of a TEM_{00} Gaussian light beam.

The spectral photon flux Φ of an FEL is the total number of photons per unit time, $\dot{N} = P_{\mathrm{FEL}}/(\hbar\omega_\ell)$, divided by the relative FEL bandwidth

$$\Phi = \frac{P_{\mathrm{FEL}}}{\hbar\omega_\ell}\left(\frac{\Delta\omega_\ell}{\omega_\ell}\right)^{-1} \quad \text{with } \Delta\omega_\ell = \sqrt{2\pi}\,\sigma_{\omega_\ell}\ . \tag{9.5}$$

Here we have approximated the almost Gaussian-shaped bandwidth curve by a rectangular bandwidth curve of equal height and area. Inserting this expression, the

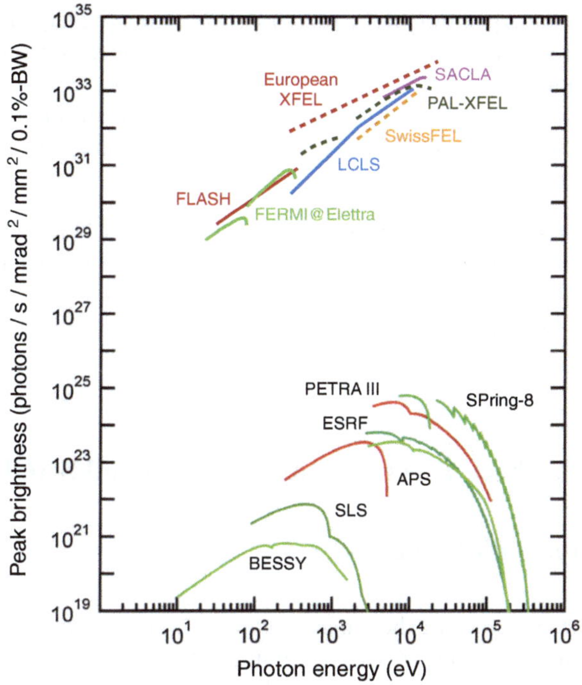

Fig. 9.1 Peak brightness (number of photons per second, mm^2, mrad^2, and 0.1 % bandwidth), plotted as a function of photon energy, of the X-ray free-electron lasers LCLS (SLAC, Stanford, USA), SACLA (RIKEN, Harima, Japan), the European XFEL (under construction in Hamburg, Germany), the SwissFEL (under construction at PSI, Villigen, Switzerland) and the PAL-XFEL (under construction at PAL, Pohang, Korea). For comparison, the peak brightness of the soft X-ray FELs FLASH (DESY, Hamburg, Germany) and FERMI@Elettra (ELETTRA, Trieste, Italy) is shown, as well as the brightness achieved at the third-generation synchrotron light sources Advanced Photon Source APS (Argonne National Lab., USA), Berliner Synchrotron BESSY (Berlin, Germany), European Synchrotron Radiation Facility ESRF (Grenoble, France), PETRA III (DESY, Hamburg, Germany), Swiss Light Source SLS (PSI, Villigen, Switzerland), and Super Photonring SPring-8 (RIKEN, Harima, Japan). We thank E. Allaria (ELETTRA), H.-D. Nuhn (SLAC), S. Reiche (PSI), H. Tanaka (RIKEN) and J.-H. Han (PAL) for providing information. The current status of FEL facilities worldwide is described in Ref. [10].

spectral photon beam brightness finally reads

$$B_{\mathrm{FEL}} = \frac{\sqrt{2}}{\pi^{3/2}\,\hbar c}\,\frac{P_{\mathrm{FEL}}}{\lambda_\ell}\,\frac{\omega_\ell}{\sigma_{\omega_\ell}}\,. \tag{9.6}$$

One can see in Fig. 9.1 that the peak brightness at LCLS and SACLA exceeds that of other accelerator-based X-ray sources by some eight orders of magnitude. Two physical reasons are responsible for the extremely high instantaneous power:

(1) the coherent superposition of the radiation fields from the large number of electrons that are present in each microbunch (this number is responsible for the large total power),
(2) the coherent superposition of the radiation fields from all microbunches within a coherence length (this effect is responsible for the small divergence and the narrow spectrum).

While the peak brightness is the essential figure of merit for a large class of important scientific applications, it should be noted there are experiments which just need a high X-ray flux. In such cases, an FEL is probably not the adequate source but a storage ring or a recirculating linac might be more appropriate. An overview is found in Ref. [11].

9.2 The X-Ray Free-Electron Lasers LCLS and SACLA

9.2.1 Layout of LCLS and SACLA

The world's first X-ray FEL, the Linac Coherent Light Source LCLS at Stanford (USA) was commissioned in 2009 and exhibited outstanding performance right from the beginning. The schematic layout of LCLS is depicted in Fig. 9.2. The electron bunches are generated in a copper photocathode by UV laser pulses from a frequency-tripled Ti:Sa laser system. The electron beam is boosted to 135 MeV and injected into the main linac. The acceleration to high energy takes place in normal-conducting traveling-wave structures L0–L3 that are made from copper and operate at 2.856 GHz.

A fourth-harmonic structure with a resonance frequency of 11.424 GHz is used for a linearization of the longitudinal phase space distribution (cf. Sect. 8.6.2). The electron bunches are longitudinally compressed in two magnetic bunch compressor chicanes (BC1 and BC2) and then accelerated to their final energy. A collimator and a magnetic deflection protect the permanent undulator magnets from radiation damage. The planar undulator magnet system is composed of 33 segments with a length of 3.4 m each. The NdFeB permanent magnets have a fixed gap of 6.8 mm and generate a field of 1.25 T. The undulator period is $\lambda_u = 30\,\text{mm}$. The magnet pole shoes are not exactly parallel but canted by an angle of 5 mrad. This allows to vary the undulator parameter K within a chosen segment by a horizontal displacement of this segment. The weak focusing produced by the undulator field is taken into account. The total length of the undulator system is 132 m, including the quadrupole magnets and diagnostic instruments between segments. The maximum active undulator length is $L_u = 112\,\text{m}$.

The world's second X-ray FEL is the "Spring-8 Angstrom Compact free-electron LAser" SACLA at Harima (Japan) which went into operation in 2012. The most remarkable feature of SACLA (see Fig. 9.3) is that sub-Ångstrøm wavelengths are reached with a comparatively modest beam energy of 8 GeV. The key component is an undulator system inside the beam vacuum chamber permitting to realize a

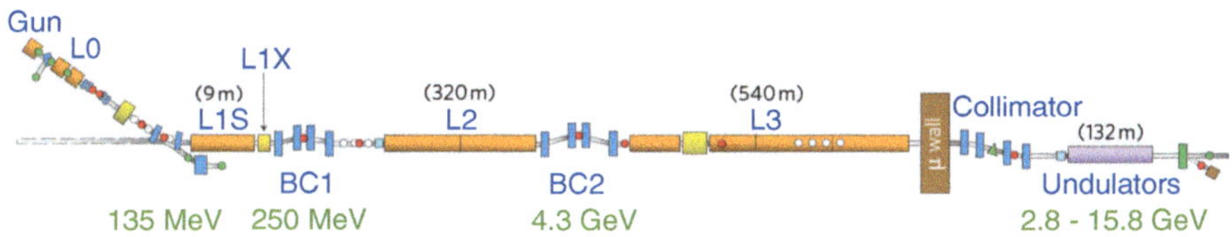

Fig. 9.2 Schematic layout of LCLS showing the electron gun, the normal-conducting accelerating structures (L0–L3), the two magnetic bunch compressor chicanes (BC1 and BC2), the collimator, and the undulators. Adapted with permission from Ref. [1]. © 2010 Macmillan Publishers Ltd.

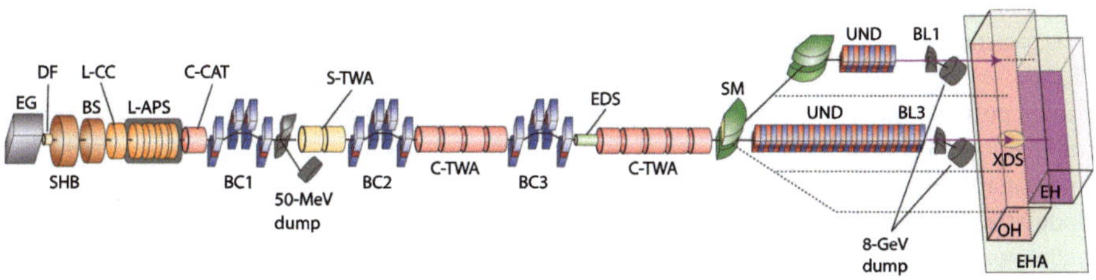

Fig. 9.3 Schematic layout of SACLA. EG: 500 keV electron gun; SHB: 238 MHz subharmonic buncher; BS: 476 MHz booster; S-TWA: S-band traveling-wave acceleration structures; C-TWA: C-band acceleration structures; BC1, BC2, BC3: bunch compressors; UND: undulators. Reprinted with permission from [2]. © 2012 Macmillan Publishers Ltd.

small gap between the magnet poles (minimum gap is 3.5 mm) and a short period of $\lambda_u = 18$ mm. The FEL wavelength can be tuned by changing the undulator parameter K through adjustment of the gap. A 5.7 GHz traveling-wave linac with a gradient of more than 35 MV/m boosts the beam energy up to 8.5 GeV. The electron source is a thermionic electron gun with a single-crystal CeB_6 cathode that generates a low-emittance beam. The electrons are accelerated by a pulsed dc voltage of 500 kV. Bunch compression is accomplished by a 238 MHz subharmonic buncher and three magnetic chicanes.

Table 9.1 summarizes the typical electron beam and FEL parameters of LCLS and SACLA. In LCLS the electron energy W can be varied between 15.8 and 2.5 GeV, corresponding to the wavelength range $1.1\,\text{Å} \leq \lambda_\ell \leq 44\,\text{Å}$. The FEL pulse energies U_{ph} are typically a few milli-Joule, the photon pulse duration T_{ph} is about 100 fs. The generation of ultrashort FEL pulses with $T_{\text{ph}} < 10$ fs is described in Sect. 9.2.2.

The excellent performance of LCLS is obvious from Fig. 9.4 where the measured power gain at 1.5 Å is plotted as a function of the active undulator length. An exponential FEL power rise is observed over four orders of magnitude. The saturation power of some 15 GW is reached after an active undulator length of 60 m which means that the total undulator length of 112 m is more than adequate for an X-ray FEL. The power gain length L_g derived from these data amounts to 3.5 m, in good agreement with the expectations according to Eq. (6.22) and a GENESIS simulation.

Table 9.1 Typical parameters of LCLS [1, 4] and SACLA (courtesy of H. Tanaka)

Parameter	Symbol	LCLS	SACLA
Electron energy	W	2.5–15.8 GeV	5.8–8.5 GeV
Peak current	I_0	0.5–3.5 kA	>3 kA
Bunch charge	Q	0.02–0.25 nC	0.2–0.3 nC
Bunch duration	T_e	<10–500 fs	20 fs
Normalized slice emittance	$\varepsilon_{n,s}$	~0.4 μm	~0.4 μm
FEL wavelength	λ_ℓ	44–1.1 Å	2.3–0.63 Å
Power gain length	L_g	~3 m	2.3 m
Saturation power	P_{sat}	3–40 GW	6–60 GW
FEL parameter	ρ_{FEL}	$\sim 5\cdot 10^{-4}$	$\sim 6.8\cdot 10^{-4}$
FEL pulse energy	U_{ph}	1–3 mJ	~0.3 mJ
Photon pulse duration	T_{ph}	<10–500 fs	2–10 fs
Undulator period	λ_u	30 mm	18 mm
Undulator field	B_0	1.25 T	1.3 T
Undulator parameter	K	3.5	≤2.2
Undulator length	L_u	112 m	90 m

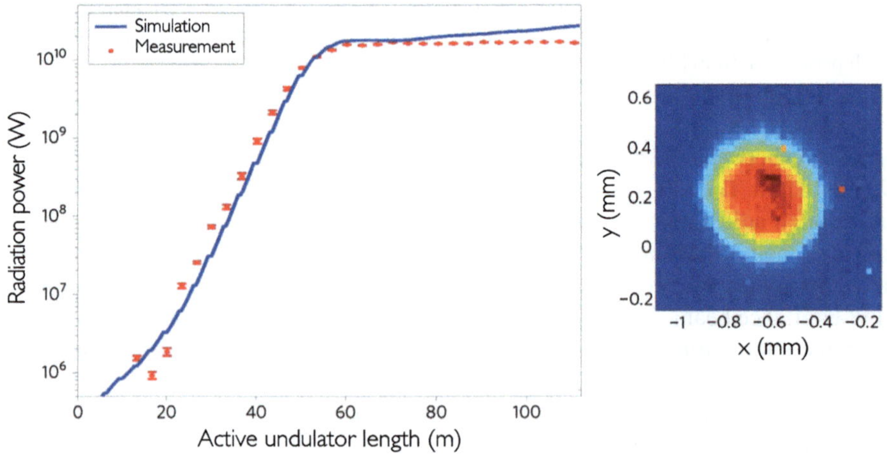

Fig. 9.4 *Left picture*: The measured power gain curve (*red* dots with error bars) of LCLS in SASE operation at the design wavelength of 1.5 Å. The solid blue curve is a simulation using the code GENESIS. *Right picture*: Image of the X-ray distribution on a scintillation screen. Figure reprinted with permission from Ref. [1]. © 2010 Macmillan Publishers Ltd.

This remarkable result is due the excellent electron beam quality. Particularly important is the small normalized slice emittance $\varepsilon_{n,s}$ of 0.4 μm. The gain curve obtained at SACLA is similarly impressive, see Fig. 9.5.

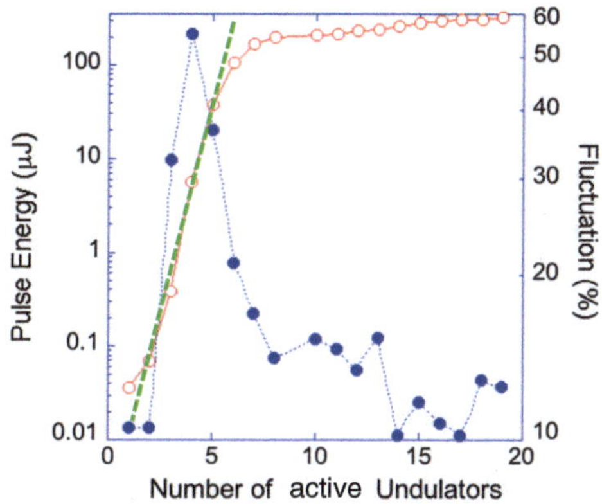

Fig. 9.5 Gain curve in SACLA at a wavelength of 1.24 Å. The FEL pulse energy (*left* ordinate) is plotted versus the number of active undulator modules together with the pulse energy fluctuation (*right* ordinate). The length of a module is 5 m. The gain length in the exponential gain region is estimated at $L_g = 2.3$ m. After reaching saturation, the pulse energy fluctuation drops below 20 %. (Courtesy of H. Tanaka).

9.2.2 Generation of Femtosecond FEL Pulses

An important task of X-ray FELs is the structural analysis of complex biomolecules and viruses by the Laue diffraction method. Many types of biomolecules cannot be arranged in a crystalline lattice, and then the diffraction pattern has to be obtained by X-ray diffraction on single molecules or on nano-meter scale crystals. Since biomolecules are typically composed of some hundred thousand hydrogen, carbon and other atoms, the X-ray pulse must contain a huge number of photons that have to be focused on a tiny target. These photons will ionize the atoms, and within a time of 20 to 50 fs the molecule undergoes a so-called Coulomb explosion. For this reason, intense X-ray pulses with a time duration below 10 fs are needed in order to obtain a complete and undisturbed diffraction pattern of the molecule before it explodes. Presently, the only known device which is capable of generating such intense and ultrashort pulses is the free-electron laser. Femtosecond FEL pulses are also useful for the study of ultrafast chemical reactions and magnetic dynamics.

The usual bunch charge of 250 pC in LCLS does not allow the longitudinal bunch compression to femtosecond-length because huge local charge densities would arise. The collective effects in the bunch compression system (space charge forces and coherent synchrotron radiation) would entirely spoil the beam quality, with the consequence of an intolerable emittance blowup and the generation of a far too large energy spread. The natural way towards ultrashort pulses is a reduction of the bunch charge by an order of magnitude. Bunch charges in the 20 pC range open the way to femtosecond FEL pulses. Additional benefits of the low-charge operation are the reduced normalized emittance and smaller wakefields. We have seen in Sect. 8.4 that space charge forces and coherent synchrotron radiation have rather detrimental

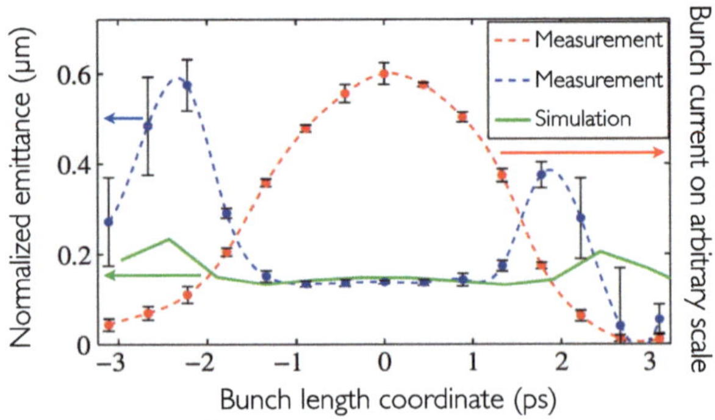

Fig. 9.6 Measurement (*blue dots* and *line*) of the normalized slice emittance of uncompressed electron bunches with the low charge of 20 pC at LCLS. The simulation (*solid green line*) is explained in [12]. The measured bunch current (*red dots* and *line*) is presented on an arbitrary scale. Figure adapted with permission from Ref. [12]. © 2009 by The American Physical Society.

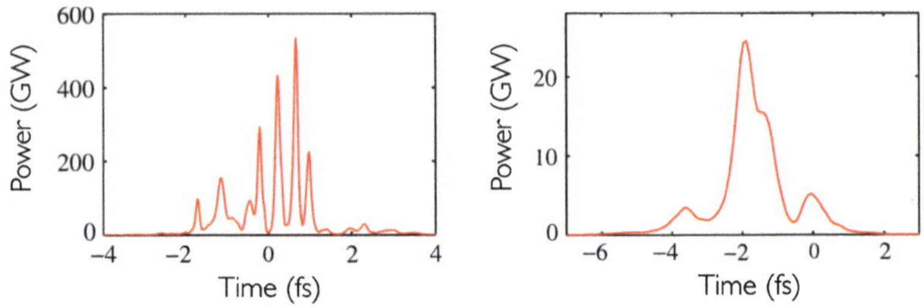

Fig. 9.7 Simulation of the time profile of FEL pulses for low-charge operation in LCLS at $\lambda_\ell =$ 0.15 nm (*left* graph) and 1.5 nm (*right* graph). The simulations were performed using GENESIS. Figure adapted with permission from Ref. [12]). © 2009 by The American Physical Society.

effects on the beam emittance. These perturbing effects are largely avoided if the bunch charge is made very low.

An operation mode with low bunch charge has been successfully established [1, 12]. For this purpose, the electron beam injector was optimized for the bunch charge of 20 pC, resulting in a very small normalized slice emittance $\varepsilon_{n,s} = 0.2\,\mu$m before bunch compression, see Fig. 9.6. These low-charge bunches were then strongly compressed and sent through the undulators where they achieved lasing. Electron beam measurements [13] confirmed the femtosecond bunch durations predicted by tracking simulations.

The computed temporal profiles of the FEL pulses at 0.15 and 1.5 nm are shown in Fig. 9.7. The two pulses have almost the same overall duration of about 2 fs, but the temporal structure is quite different. The 0.15 nm FEL pulse has clearly more wave packets (longitudinal modes) than the 1.5 nm pulse, which is a consequence of the shorter coherence time of a few hundred attoseconds (cf. Eq. 7.15). Experimentally, an FEL pulse duration of about 2 fs was determined by cross-correlation [14].

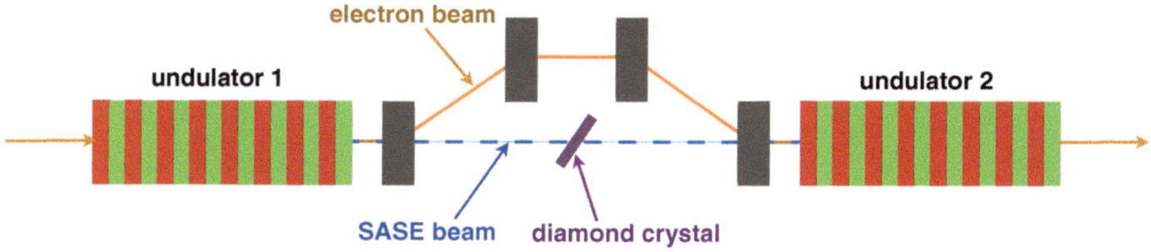

Fig. 9.8 Schematic of the self-seeding setup at LCLS. To switch from seeding operation to SASE operation the diamond crystal is retracted and the magnetic chicane is switched off.

9.2.3 Monochromatic X-Rays from Self-Seeding

We have seen that self-amplified spontaneous emission is a very powerful mechanism for generating laser light in a wavelength regime where conventional lasers are unavailable or have a poor performance. SASE FELs are capable of producing X-ray pulses with a power of some 10 GW and a time duration in the 10 fs range. The transverse coherence is excellent but the longitudinal coherence (identical with the temporal coherence) is poor. As discussed in Chap. 7, SASE radiation has a fairly wide spectrum with strong wavelength and intensity fluctuations from pulse to pulse. For spectroscopic experiments, an X-ray source with better monochromaticity would be highly desirable. Significant progress in this direction can be expected if the lasing process is not initiated by the statistically fluctuating undulator radiation but rather by the nearly monochromatic light wave of a seed laser, see Sect. 7.4. For X-rays, suitable seed lasers are presently not available. As a practical alternative, the idea of self-seeding was proposed at DESY [15–17] which works as follows. SASE radiation is produced in a first undulator section and then passed through a monochromator. A magnetic bypass guides the electron beam around the monochromator. The path-length traveled by the electrons is adjusted such that the emerging electron bunch and the monochromatized radiation pulse have a good temporal overlap in the second undulator section in order to achieve a high FEL gain. The microbunching acquired in the first undulator section during the SASE process is washed out in the magnetic bypass, and hence it does not disturb the seeding process in the second undulator section.

A successful self-seeding experiment has been carried out at LCLS [18]. The experimental setup is shown schematically in Fig. 9.8. The first undulator section comprises 15 modules of 3.4 m length, 13 of which are used to generate SASE radiation with an average power of 1 GW. The radiation is passed through a special diamond crystal in the so-called *forward Bragg diffraction geometry* [18], where a transmitted X-ray pulse with a relatively long monochromatic tail is generated. The first maximum of this wake has a small delay of only 19 fs with respect to the incident SASE pulse. A magnetic chicane with a parallel beam displacement of just 2.5 mm is sufficient to delay the electron bunch by these 19 fs, to guide it around the diamond crystal and to bring it into overlap with the radiation tail, thereby selecting

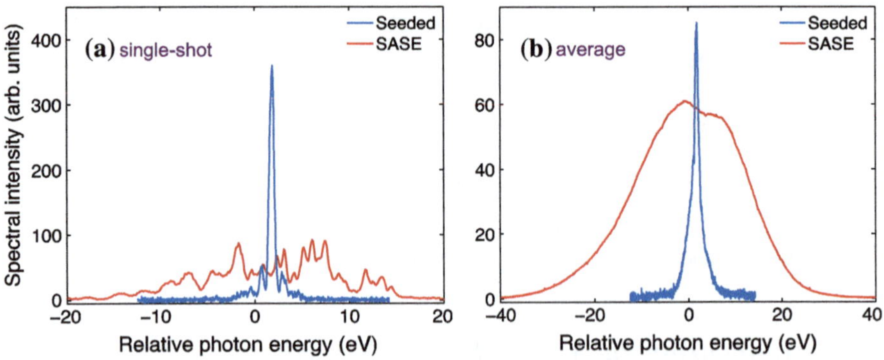

Fig. 9.9 Comparison of the measured X-ray spectra for the self-seeded FEL mode of operation (*blue*) and the SASE mode (*red*). **a** Single-shot measurement of a self-seeding pulse of large amplitude in comparison with a single-shot SASE pulse. **b** Comparison of the averaged X-ray spectra for the self-seeding and SASE modes. The central photon energy is 8.3 keV. Note the different horizontal and vertical scales in (**a**) and (**b**). (Courtesy of A. Lutman and Z. Huang).

a monochromatic wave for seeding. FEL seeding happens in the second undulator section consisting of 13 modules with a total length of 52 m.

Measurements in the self-seeding mode and the SASE mode are compared in Fig. 9.9. In single-shot operation the seeded pulse consists of a single narrow spike whereas the SASE pulse is composed of many narrow spikes scattered over a rather wide range of photon energies. Averaged over many shots the seeded radiation has a bandwidth of 0.4 eV (FWHM) while the SASE radiation has a bandwidth of 20 eV, so a bandwidth reduction by a factor of about 50 is achieved by self-seeding.

The single-shot measurement shown in this figure is a favourable case, the seeded-FEL pulse has a much higher peak intensity than the SASE spikes. The average seed signal is lower, as can be seen from Fig. 9.9b. In fact, the pulse energy stability of the seeded FEL is still poor. Large fluctuations occur that can be traced back to corresponding energy fluctuations in the SASE pulses used for seeding. The length of the first undulator section must be chosen such that the SASE FEL process stays below saturation,[1] and consequently the seed power fluctuates close to 100 % because a single longitudinal mode is selected (compare Sect. 7.3.3 and Fig. 7.14). The second half of the undulator is presently shorter than the saturation length, so the seeded FEL process remains in the linear regime. The energy jitter of the seed pulse is thus translated 1:1 to the output pulse. There is room for improvement though, for example by adding undulator modules to the second undulator section in order to achieve saturation in the seeded FEL process. We have seen in Fig. 5.5 that the saturation power of a seeded FEL does not depend on the power of the seed beam.

[1] Making the first undulator section longer than the SASE saturation length has a serious disadvantage: the electron bunches acquire a large energy spread in the saturated SASE process which in turn impedes the seeded FEL gain in the second undulator section.

The great potential offered by self-seeding is obvious from these results. Considerable progress can be expected in the near future.

9.2.4 Electron Beam Energy Loss in the Undulator Magnets

When an electron bunch with N_e electrons passes through the long undulator of an X-ray FEL, not only the desired FEL radiation is emitted but in addition a large amount of spontaneous undulator radiation. The two types of radiation have quite a different dependence on the electron energy. According to Eq. (2.21), the spontaneous radiation power emitted by an electron

$$P_{\rm rad} = \frac{e^4 \gamma^2 B_0^2}{12\pi\varepsilon_0 c\, m_e^2}$$

grows quadratically with the electron energy $W = \gamma m_e c^2$. In contrast to this, the FEL saturation power (6.27)

$$P_{\rm sat} \approx 1.6\, \rho_{\rm FEL}\, P_{\rm beam} \left(\frac{L_{g0}}{L_g}\right)^2$$

increases only rather weakly with energy. The electron beam power $P_{\rm beam}$ scales linearly with energy but the FEL parameter drops by about a factor of 7 if one compares FLASH ($W = 1$ GeV) and LCLS ($W = 13.6$ GeV). Hence one can expect that beyond some energy threshold the electron beam loses more energy by spontaneous undulator radiation than by the FEL gain process. A rough estimate is obtained as follows. The energy carried away by spontaneous radiation in an undulator magnet of length L_u is

$$U_{\rm spon} = N_e P_{\rm rad} T_{\rm tran}\,,$$

where $T_{\rm tran} = L_u/(\bar{\beta}c) \approx L_u/c$ is the transit time of the electron bunch through the undulator, which amounts to 373 ns for the 112 m long undulator of LCLS.[2] The energy transferred to the FEL light pulse is

$$U_{\rm FEL} \approx P_{\rm sat} T_{\rm ph}\,.$$

In this case, however, $T_{\rm ph}$ is the time duration of the FEL photon pulse which is very short and in the 10–100 fs regime. For typical LCLS parameters one finds

$$\frac{U_{\rm spon}}{W_{\rm bunch}} = 1.1 \cdot 10^{-3}\,, \quad \frac{U_{\rm FEL}}{W_{\rm bunch}} = 0.7 \cdot 10^{-3}\,,$$

[2] Note that the electron transit time has nothing to do with the time duration of the spontaneous radiation pulse which is in 10–100 fs range just like the FEL pulse.

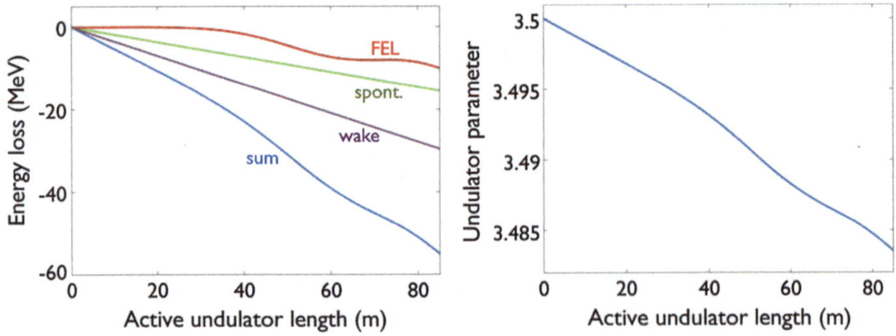

Fig. 9.10 *Left*: Calculated energy loss of a 13.6 GeV electron beam as a function of undulator length. The loss is due to the FEL gain process (*red*), spontaneous undulator radiation (*green*) and resistive-wall wake fields (*purple*). *Right*: Variation of the undulator parameter K that is needed to keep the light wavelength λ_ℓ constant, thereby preserving the resonance condition.

where $W_{\text{bunch}} = N_e W$ is the sum of the kinetic energies of the electrons in the bunch. Resistive-wall wake fields constitute another loss mechanism which is indeed the dominating effect at LCLS. The calculated energy loss is plotted in Fig. 9.10 as a function of the path length traveled in the undulator.

When the electrons travel through the long undulator system, their fractional energy loss will eventually exceed the energy bandwidth of LCLS, which is given by the FEL parameter $\rho_{\text{FEL}} = 5 \cdot 10^{-4}$. One can say that the electrons "fall out of resonance". In order to restore the resonance condition one can gradually reduce the undulator parameter with increasing z by reducing the magnetic field. At LCLS this field reduction is easy to accomplish owing to the canted undulator pole shoes mentioned before. A slight horizontal movement of the undulator sections in question is sufficient. This correction method is often called "undulator tapering", and by applying undulator tapering the saturation power can be more than doubled.

9.3 Undulator Radiation Background and Quantum Effects

9.3.1 Background from Undulator Radiation

In view of the big amount of spontaneous undulator radiation energy, the question arises what the benefit of FEL radiation is at all in the X-ray regime. This benefit is in fact present and it is overwhelming. The superiority of FEL radiation results from the fact that it is emitted into a tiny solid angle and a narrow spectral band, while undulator radiation has a much wider angular distribution and moreover exhibits a strong wavelength variation with increasing emission angle.

The solid angle subtended by FEL radiation can be estimated as follows. According to Sect. 9.1 the FEL beam in the saturation regime is well described by the

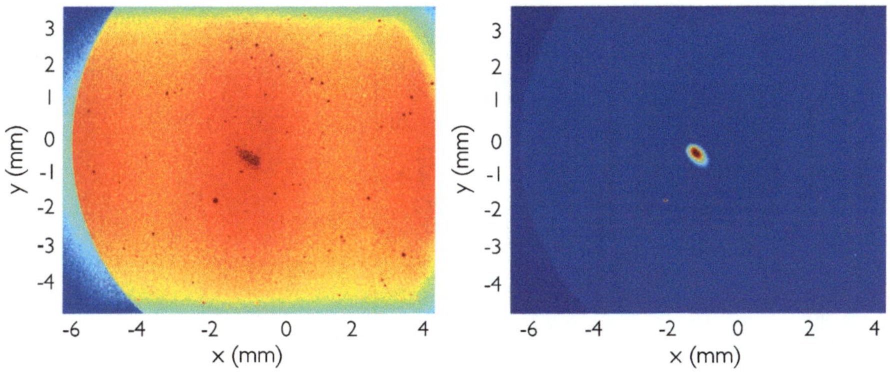

Fig. 9.11 Transverse photon beam profiles measured with a Ce:YAG scintillation screen 50 m downstream of the LCLS undulators. *Left picture*: Low FEL gain. 10 undulator segments are active, the peak current is 0.5 kA. The picture is dominated by undulator radiation, the weak FEL radiation appears as a dark spot in the center. *Right picture*: High FEL gain, 12 undulator segments are active and the peak current is 3.0 kA. The sensitivity of the pixel camera has been reduced to avoid overexposure of the bright FEL spot, therefore undulator radiation is no longer visible. (Courtesy of P. Emma and Z. Huang).

fundamental Gaussian mode. It is shown in Sect. 10.3 that the rms divergence angle of a TEM_{00} beam is $\sigma_\theta = \lambda_\ell/(4\pi\sigma_x)$. If the photon beam is matched in size to the electron beam, $\sigma_x = \sqrt{\varepsilon_x\beta_x}$, the divergence angle at LCLS is $\sigma_\theta \approx 3\cdot 10^{-7}$rad, and the corresponding solid angle is

$$\Delta\Omega_{\mathrm{FEL}} < 10^{-12}\ \mathrm{sterad}\,. \tag{9.7}$$

In contrast to this tiny angular divergence, the opening cone angle (2.27) containing the major fraction of spontaneous undulator radiation power is significantly larger[3] with

$$\theta_{\mathrm{spont}} = \frac{K}{\gamma} \sim 10^{-4}\ \mathrm{rad}\,, \quad \Delta\Omega_{\mathrm{spon}} \approx 10^{-7}\ \mathrm{sterad}\,.$$

The corresponding solid angle is some five orders of magnitude larger than the solid angle of the FEL radiation, so there will be a strong suppression of the spontaneous radiation if one restricts the aperture of the detector to the small solid angle $\Delta\Omega_{\mathrm{FEL}}$. Figure 9.11 is an illustration of these considerations. The photon distribution in a plane transverse to the photon beam direction was measured with a scintillation screen 50 m downstream of the LCLS undulators. The left image was made at low FEL gain and is dominated by the large background from spontaneous undulator radiation; the onset of the FEL gain process is barely visible as a small dark spot at the center. The

[3] The rather tight collimation of the first harmonic of undulator radiation given in Eq. (2.26) is due to the requirement that the angle-dependent wavelength shift stays within the spectral bandwidth observed in forward direction. Dropping this requirement a large amount of radiation is found at substantially larger angles.

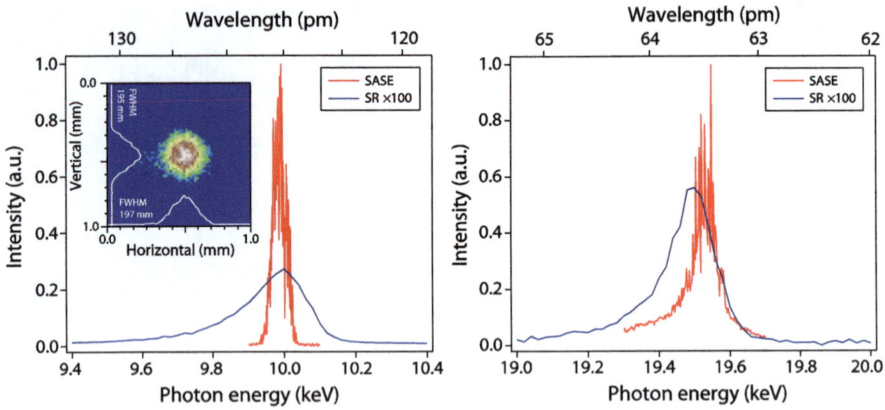

Fig. 9.12 Measured spectra of SASE-FEL radiation (*red curves*) and spontaneous radiation (*blue curves*) at $\lambda_\ell = 1.24\,\text{Å}$ (*left*) and at $\lambda_\ell = 0.635\,\text{Å}$ (*right*). Data are from SACLA. Figure reprinted with permission from [2]. © 2012 by Macmillan Publishers Ltd.

right image was taken at high FEL gain, but with a much reduced sensitivity of the pixel camera. Here only the bright spot from FEL radiation is visible.

Likewise, the spectral bandwidth of SASE FEL radiation is narrow at saturation:

$$\left(\frac{\Delta_{\omega_\ell}}{\omega_\ell}\right) \approx 2.5\,\rho_{\text{FEL}} \approx 10^{-3}\,,$$

while the undulator radiation entering a small-aperture spectrometer has a wider spectrum and a much lower intensity. This is demonstrated in Fig. 9.12 with data from SACLA. The measured spectra of SASE-FEL radiation and of spontaneous radiation (the latter one multiplied with 100) are shown.

9.3.2 Quantum Effects and Beam Energy Spread

The classical FEL theory turns out to be adequate for a description of all existing and proposed X-ray FEL facilities. Nevertheless, the quantum nature of the photons, which plays an essential role for synchrotron radiation in storage rings, must be taken into account. When an electron in a storage ring emits a photon, its energy will abruptly change, and this *quantum recoil* affects the beam dynamics by exciting a horizontal betatron oscillation. In an FEL, the emission of a single photon has a much smaller impact. It is shown in Ref. [19] that quantum effects become only relevant in a high-gain FEL when the fractional energy loss due to single-photon emission is comparable to the FEL bandwidth:

$$\frac{\hbar\omega_\ell}{W} \simeq \rho_{\text{FEL}} \, .$$

In the case of LCLS with an electron beam energy of $W = 13.6\,\text{GeV}$ and a photon energy of $\hbar\omega_\ell = 8.27\,\text{keV}$, the fractional energy loss of an electron is only $6 \cdot 10^{-7}$ and thus far below the FEL parameter $\rho_{\text{FEL}} \approx 5 \cdot 10^{-4}$. Hence, quantum recoil is of no importance as long as only single-photon emission is considered. However, the accumulated beam energy loss due to the many photons that are emitted in the long undulator of an X-ray FEL may exceed the FEL bandwidth, as discussed above.

Spontaneous undulator radiation has another undesirable effect. An additional energy spread σ_γ is induced within the electron beam that has its origin in the quantum nature of the radiation. The growth rate of this additional energy spread σ_γ is [20, 21]:

$$\frac{d\sigma_\gamma^2}{dz} = \frac{7e^2\hbar\gamma^4 K^3 k_u^3}{60\pi\varepsilon_0 m_e^2 c^3}\left(1.2 + \frac{1}{K + 1.33\,K^2 + 0.4\,K^3}\right). \tag{9.8}$$

For the 13.6 GeV electron beam at the LCLS, the additional energy spread after 60 m of undulators is $\sigma_\gamma/\gamma \approx 0.2\,\rho_{\text{FEL}}$. The energy spread caused by this so-called *quantum diffusion* effect is thus of minor importance. The growth rate (9.8) increases with the fourth power of the electron energy. Quantum diffusion will thus be a serious concern for X-ray FELs with significantly higher electron beam energies than in LCLS.

The quantum regime of FELs starting from noise is discussed in Ref. [22].

9.4 X-Ray Beam Lines

The peak intensity of the X-ray beam leaving the undulator is many orders of magnitude higher than what can currently be handled with existing optical technologies. The safest method of reducing the power density without degrading the beam quality is to let the photon beam propagate through a long vacuum pipe until it is spread out over a substantially increased area via its natural divergence. Thereby a tolerable irradiation level for optical elements like mirrors, lenses, and monochromators can be achieved.

What matters for the optical elements of the photon beam lines of an X-ray FEL is the high instantaneous energy dose deposited in a thin surface layer. During the femtosecond X-ray pulse the surface layer is far away from thermal equilibrium, and hence it is meaningless to consider equilibrium properties like melting temperature or expansion coefficient. From experience, the tolerable energy dose causing no damage is approximately 0.01 eV/atom, much less than the energy required for melting (about 0.9 eV/atom for graphite).

In order to reduce the energy dose, the optical elements are designed to achieve the highest reflectivity possible. For mirrors this can be realized by choosing a grazing incidence geometry with angles below the critical angle for total external reflection (remember that the refractive index of common materials is slightly less than 1 in

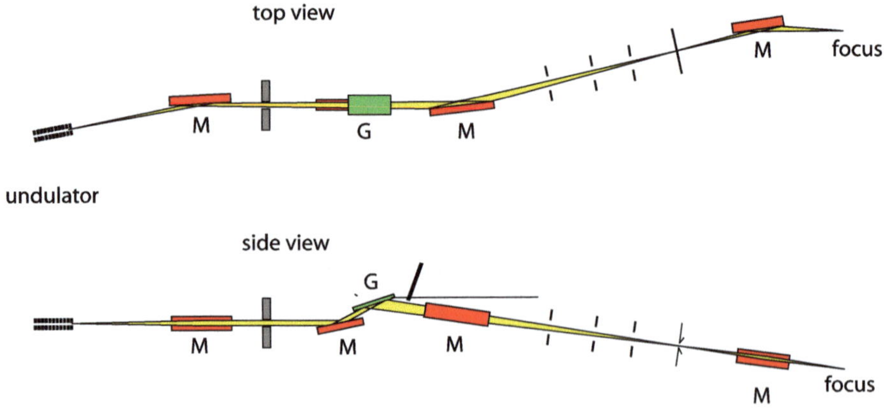

Fig. 9.13 *Top view* and *side view* of a 400 m long X-ray beam line equipped with a high-power grating monochromator, foreseen for the European XFEL. M are grazing-incidence mirrors, G is a diffraction grating. (Courtesy of Th. Tschentscher).

the X-ray regime). This is the only way for reaching a reflectivity of more than 90 % over a wide range of X-ray wavelengths. A good material combination is a highly polished silicon substrate with carbon coating. For Ångstrøm radiation, such mirrors are operated at a glancing angle of 2 mrad, but even then photon beam drift lengths of some 100 m are needed to stay below the 0.01 eV/atom limit. The consequence of the grazing incidence geometry and the long drift length is that the mirrors and lenses have to be about 0.5 m long. The surface must be polished to 0.1 nm residual roughness and 0.3 μrad tangential slope errors to preserve the wave fronts of the FEL radiation. As an example, a photon beam line design for the European XFEL is shown in Fig. 9.13.

Many experiments at X-ray FELs cannot cope with the maximum power of the FEL. While is it always possible to reduce the X-ray intensity, for example by choosing a low bunch charge, in practice many experiments with different intensity requirements will be running in parallel, and then a variable power reduction in a specific beamline may be needed. At LCLS the photon beam attenuation is accomplished by the system shown in Fig. 9.14. The main components are beryllium disks with a thickness between 0.1 and 32 mm, which can be inserted into the photon beam, and a nitrogen gas absorption cell. Differential pumping sections isolate the gas cell from the ultra-high vacuum in the upstream and downstream sections of the photon beam line. The reduction of the photon pulse energy can be measured with two gas detectors which are installed upstream and downstream of the attenuation section.

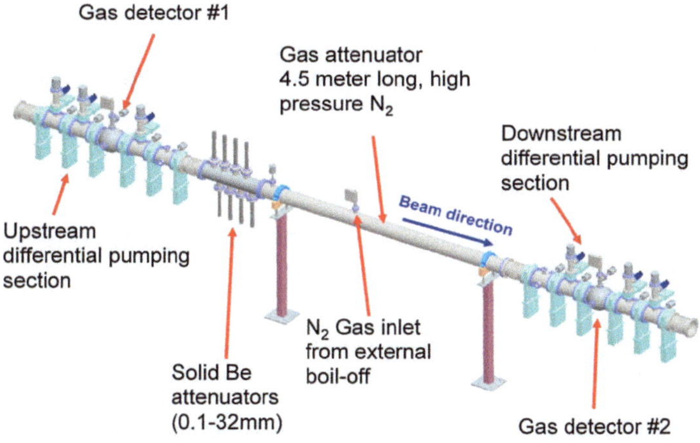

Fig. 9.14 Photon beam attenuation system at the LCLS, consisting of beryllium (Be) disks of different thickness and a nitrogen (N_2) gas absorption cell. The attenuation system is located between two differential pumping sections. These incorporate two gas detectors for measuring the FEL pulse energy before and after the attenuation. Figure reprinted with permission from Ref. [23]. © 2012 Elsevier BV.

References

1. P. Emma et al., First lasing and operation of an ångstrom-wavelength free-electron laser. Nat. Photonics **4**, 641 (2010)
2. T. Ishikawa et al., A compact X-ray free-electron laser emitting in the sub-ångström region. Nat. Photonics **6**, 540 (2012)
3. R. Colella, A. Luccio, Proposal for a free electron laser in the X-ray region. Opt. Commun. **50**, 41 (1984)
4. J. Arthur et al., Linac Coherent Light Source (LCLS) Conceptual Design Report, Report SLAC-R-593, Menlo Park (2002)
5. R. Brinkmann et al., XFEL: The European X-Ray Free-Electron Laser. Technical design report, Report DESY-06-097, Hamburg (2006)
6. Z. Huang, K.-J. Kim, Review of x-ray free-electron laser theory. Phys. Rev. ST Accel. Beams **10**, 034801 (2007)
7. H. Wiedemann, *Synchrotron Radiation* (Springer-Verlag, Berlin, 2003)
8. W. Ackermann et al., Operation of a free-electron laser from the extreme ultraviolet to the water window. Nat. Photonics **1**, 336 (2007)
9. E. Allaria et al., Highly coherent and stable pulses from the FERMI seeded free-electron laser in the extreme ultraviolet. Nat. Photonics **6**, 699 (2012)
10. http://www.lightsources.org/
11. U. Bergmann et al., Science and Technology of Future Light Sources: A White Paper, Reports ANL-08/39, BNL-81895-2008, LBNL-1090E-2009, and SLAC-R-917 (2008)
12. Y. Ding et al., Measurements and Simulations of Ultralow Emittance and Ultrashort Electron Beams in the Linac Coherent Light Source. Phys. Rev. Lett. **102**, 254801 (2009)
13. Z. Huang et al., Measurement of femtosecond LCLS bunches using the SLAC A-Line spectrometer, in *Proceedings Proceedings of the 24th Particle Accelerator Conference*, New York, USA, THP183 2011
14. Y. Ding et al., Femtosecond X-ray pulse characterization in free-electron lasers using a cross-correlation technique. Phys. Rev. Lett. **109**, 254802 (2012)

15. J. Feldhaus et al., Possible application of X-ray optical elements for reducing the spectral bandwidth of an X-ray SASE FEL. Opt. Commun. **140**, 341 (1997)
16. E.L. Saldin, E.A. Schneidmiller, Yu. Shvyd'ko, M.V. Yurkov, X-ray FEL with a meV bandwidth. Nucl. Instrum. Meth. A **475**, 357 (2001)
17. G. Geloni, V. Kocharyan, E.L. Saldin, A novel self-seeding scheme for hard X-ray FELs. J. Mod. Opt. **38**, 1391–1403 (2011)
18. J. Amann et al., Demonstration of self-seeding in a hard-X-ray free-electron laser. Nat. Photonics **6**, 693–698 (2012)
19. C.B. Schroeder, C. Pellegrini, P. Chen, Quantum effects in high-gain free-electron lasers. Phys. Rev. E **64**, 056502 (2001)
20. M. Sands, The Physics of Electron Storage Rings: An Introduction, Report SLAC-121, Menlo Park (1970)
21. J. Rossbach, E.L. Saldin, E.A. Schneidmiller, M.V. Yurkov, Fundamental limitations of X-ray FEL operation due to quantum fluctuations of undulator radiation. Nucl. Instrum. Meth. A **393**, 152 (1997)
22. R. Bonifacio, N. Piovella, G.R.M. Robb, A. Schiavi, Quantum regime of free electron lasers starting from noise. Phys. Rev. ST Accel. Beams **9**, 090701 (2008)
23. S. Moeller et al., Photon beamlines and diagnostics at LCLS. Nucl. Instrum. Meth. A **635**, 6 (2010)

Chapter 10
Appendices

10.1 Hamiltonian Formalism

10.1.1 Basic Elements of the Hamiltonian Formalism

In this section we demonstrate how Hamiltonian mechanics can be used to derive the trajectory of an electron in an undulator magnet and its coupling to the radiation field. For a thorough presentation of the Lagrange–Hamilton formulation of classical mechanics we refer to the textbooks by Landau and Lifshitz [1] and Goldstein [2].

10.1.1.1 Non-relativistic Hamiltonian

In non-relativistic mechanics the Hamilton function of a particle is the sum of its kinetic and potential energies

$$H(q_j, p_j, t) = W_{\text{kin}} + W_{\text{pot}}. \tag{10.1}$$

The q_j and p_j are the generalized coordinates and momenta. The Hamilton equations express the time derivatives of the coordinates and momenta in terms of partial derivatives of the Hamiltonian

$$\dot{q}_i = \frac{\partial H}{\partial p_i}, \quad \dot{p}_i = -\frac{\partial H}{\partial q_i}. \tag{10.2}$$

The total time derivative of the Hamiltonian is

$$\frac{dH}{dt} = \frac{\partial H}{\partial t} + \sum_j \left[\frac{\partial H}{\partial q_j} \dot{q}_j + \frac{\partial H}{\partial p_j} \dot{p}_j \right]. \tag{10.3}$$

Because of the Hamilton equations (10.2) the second term vanishes, hence

P. Schmüser et al., *Free-Electron Lasers in the Ultraviolet and X-Ray Regime*, Springer Tracts in Modern Physics 258, DOI: 10.1007/978-3-319-04081-3_10,

$$\frac{dH}{dt} = \frac{\partial H}{\partial t}. \tag{10.4}$$

This means that the total energy of the particle is conserved if the Hamilton function has no explicit time dependence. The implicit time dependence contained in the generalized coordinates and momenta may increase the kinetic energy at the expense of the potential energy, or vice versa, but the sum of both remains invariant.

10.1.1.2 Example: Mathematical Pendulum

A mechanical example which is of relevance for the FEL is the mathematical pendulum. The mathematical pendulum consists of a mass m attached to a massless bar of length ℓ. In this case the natural choice for the canonical coordinate is the angle ϕ, the conjugated canonical momentum is then the angular momentum L

$$q = \phi, \quad p = L = m\,\ell^2\,\dot{\phi} = m\,\ell^2\,\omega.$$

The kinetic and potential energies and the Hamiltonian are

$$\begin{aligned} W_{\text{kin}} &= \frac{L^2}{2m\ell^2}, \quad W_{\text{pot}} = m\,g\,\ell(1 - \cos\phi), \\ H(\phi, L) &= \frac{L^2}{2m\ell^2} + m\,g\,\ell(1 - \cos\phi), \end{aligned} \tag{10.5}$$

where g is the acceleration of gravity. The Hamiltonian is independent of time, and hence the total energy is conserved:

$$W = W_{\text{kin}} + W_{\text{pot}} = \frac{L^2}{2m\ell^2} + m\,g\,\ell(1 - \cos\phi) = \text{const.}$$

The Hamilton equations are

$$\frac{d\phi}{dt} = \frac{\partial H}{\partial L} = \frac{L}{m\ell^2}, \quad \frac{dL}{dt} = -\frac{\partial H}{\partial \phi} = -m\,g\,\ell\,\sin\phi. \tag{10.6}$$

The trajectories in the phase space (ϕ, L) can be easily constructed by writing the coupled differential equations (10.6) as difference equations and solving these in small time steps. These trajectories are the curves of constant Hamiltonian. For small angles we have $\sin\phi \approx \phi$, and the pendulum carries out a harmonic oscillation of the form:

$$\phi(t) = \phi_0 \cos(\omega t), \quad L(t) = -m\,\ell^2\omega\,\phi_0 \sin(\omega t),$$

corresponding to an elliptic phase-space curve. With increasing angular momentum the motion becomes anharmonic. At very large angular momentum one gets a rotation

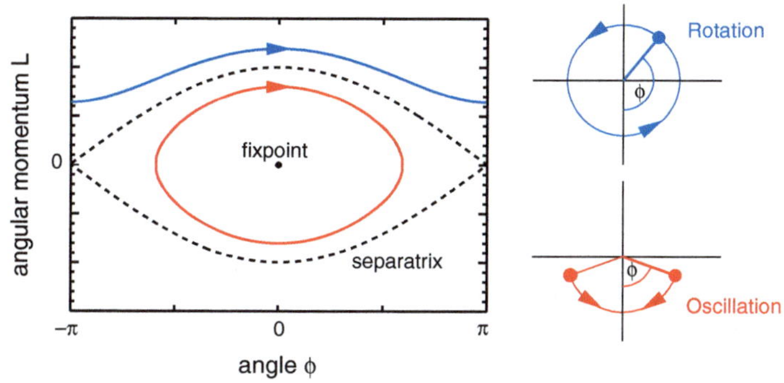

Fig. 10.1 Phase space curves of a mathematical pendulum. The separatrix (*dashed curve*) separates the region of bounded motion (*periodic oscillation*) from that of unbounded motion (*rotation*). The point (0, 0) is a fixpoint.

(unbounded motion). The two regions of motion are separated by a curve called the separatrix. The equation of the separatrix can be derived from the initial conditions

$$\phi_0 = \pi, \quad L_0 = 0 \Rightarrow H_{\text{sep}} = m\,g\,\ell(1 - \cos\phi_0) = 2\,m\,g\,\ell.$$

Since the Hamiltonian is constant on the separatrix, the angular momentum as a function of the angle ϕ can be computed from

$$\frac{[L_{\text{sep}}(\phi)]^2}{2m\ell^2} = H_{\text{sep}} - m\,g\,\ell(1 - \cos\phi)$$

from which follows

$$L_{\text{sep}}(\phi) = \pm 2m\ell^2\sqrt{g/\ell}\,\cos(\phi/2). \tag{10.7}$$

The phase space picture of the mathematical pendulum is shown in Fig. 10.1.

10.1.1.3 Relativistic Hamiltonian

The relativistic Hamiltonian of a particle of rest mass m_0 moving in a force- and field-free region is

$$H(q_i, p_i) \equiv \gamma m_0 c^2 = c\sqrt{\boldsymbol{p}^2 + m_0^2 c^2} = c\left[\boldsymbol{p}^2 + m_0^2 c^2\right]^{1/2}. \tag{10.8}$$

The canonical coordinates q_i are taken here as the Cartesian coordinates $\boldsymbol{r} = (x, y, z)$, and the canonical momentum is $\boldsymbol{p} = \gamma m_0 \boldsymbol{v}$. The Hamiltonian is identical with the total relativistic energy of the particle, for a free particle it is the sum of kinetic energy W_{kin} and rest energy $m_0 c^2$.

In the presence of an electromagnetic field the Hamiltonian of a particle with charge q must be modified. We characterize the field by its scalar and vector potentials:

$$\boldsymbol{E} = -\nabla \Phi - \frac{\partial \boldsymbol{A}}{\partial t}, \quad \boldsymbol{B} = \nabla \times \boldsymbol{A}. \tag{10.9}$$

Remember that we use the SI system in contrast to many articles and books on FEL theory that use Gaussian cgs units. To account for the electromagnetic field we have to add the potential energy $W_{\rm pot} = q\,\Phi$, and moreover the kinetic momentum $\boldsymbol{p} = \gamma m_0 \boldsymbol{v}$ has to be replaced by the *canonical momentum*

$$\boldsymbol{P} = \boldsymbol{p} + q\boldsymbol{A} = \gamma m_0 \boldsymbol{v} + q\boldsymbol{A}. \tag{10.10}$$

The relativistic Hamiltonian of a charged particle in an electromagnetic field is thus

$$H(\boldsymbol{r}, \boldsymbol{P}, t) = c\left[(\boldsymbol{P} - q\boldsymbol{A})^2 + m_0^2 c^2\right]^{1/2} + q\,\Phi. \tag{10.11}$$

Again this Hamiltonian is identical with the total relativistic energy of the particle, namely the sum of kinetic energy, potential energy, and rest energy $m_0 c^2$. The Hamiltonian has to be considered a function of the canonical coordinates $\boldsymbol{r} = (x, y, z)$, the canonical momenta $\boldsymbol{P} = (P_x, P_y, P_z)$, and the time t.

In the following we consider an electron (charge $q = -e$, rest mass $m_0 = m_e$) moving in the magnetic field of an undulator magnet and in the electric field of a light wave. Then no scalar potential Φ is present and the Hamilton function has the form

$$H(\boldsymbol{r}, \boldsymbol{P}, t) = c\left[(\boldsymbol{P} + e\boldsymbol{A})^2 + m_e^2 c^2\right]^{1/2} \equiv \gamma m_e c^2. \tag{10.12}$$

The vector potential $\boldsymbol{A}(\boldsymbol{r}, t)$ comprises both fields. This Hamiltonian is equal to the total energy $W = \gamma m_e c^2$ of the electron and will be time-dependent if the electron exchanges energy with the light wave, in which case the Lorentz factor is a function of time, too. The Hamilton equations are evaluated using the formula

$$\frac{d\sqrt{f(x)}}{dx} = \frac{f'(x)}{2\sqrt{f(x)}}.$$

With the help of this formula we find for example

$$\frac{dx}{dt} = \frac{\partial H}{\partial P_x} = \frac{2c(P_x + eA_x)}{2\sqrt{(\boldsymbol{P} + e\boldsymbol{A})^2 + m_e^2 c^2}} = \frac{(P_x + eA_x)}{\gamma m_e} = \frac{\gamma m_e v_x}{\gamma m_e} = v_x. \tag{10.13}$$

Note that one can replace the square root in the denominator by $\gamma m_e c$. The time derivative of the Hamiltonian is found by the same method

$$\frac{\partial H}{\partial t} = \frac{(\boldsymbol{P} + e\boldsymbol{A})}{\gamma m_e} \cdot \frac{\partial(e\boldsymbol{A})}{\partial t} = -e\,\boldsymbol{v} \cdot \boldsymbol{E}, \tag{10.14}$$

where we have used $\boldsymbol{P} + e\boldsymbol{A} = \gamma m_e \boldsymbol{v}$ and $\boldsymbol{E} = -\partial \boldsymbol{A}/\partial t$.

10.1.2 Electron Motion in a Planar Undulator

The Hamiltonian formulation is now applied to compute the electron trajectory in a planar undulator, following partly the procedure in Ref. [3]. The emission of undulator radiation is neglected for the time being. We consider a simplified form of the undulator field, see Eq. (2.3)

$$\boldsymbol{B} = -B_0 \sin(k_u z)\boldsymbol{e}_y \tag{10.15}$$

which can be derived from the vector potential

$$\boldsymbol{A} = \frac{B_0}{k_u} \cos(k_u z)\boldsymbol{e}_x. \tag{10.16}$$

Here $\boldsymbol{e}_x$ and $\boldsymbol{e}_y$ are the unit vectors in x and y direction, respectively. An electric field is not present in the undulator, hence the potential energy term is missing. We note that the vector potential has only an x component which depends on the single variable z. The Hamiltonian of the electron becomes thus

$$H(z, P_x, P_y, P_z) = c\left[\left(P_x + e\frac{B_0}{k_u}\cos(k_u z)\right)^2 + P_y^2 + P_z^2 + m_e^2 c^2\right]^{1/2}$$
$$\equiv \gamma m_e c^2. \tag{10.17}$$

The two Hamilton equations

$$\dot{P}_x = -\frac{\partial H}{\partial x} = 0, \quad \dot{P}_y = -\frac{\partial H}{\partial y} = 0$$

imply that $P_x = \gamma m_e v_x - eA_x$ and $P_y = \gamma m_e v_y$ are constants of motion. The vector potential A_x vanishes at $z = \lambda_u/4$, right after the beginning of the undulator. Choosing the initial conditions such that $v_x = v_y = 0$ at this position, we obtain $P_x, P_y \equiv 0$ all along the undulator from which follows $v_y(z) \equiv 0$ and

$$v_x(z) = \frac{eB_0}{\gamma m_e k_u}\cos(k_u z) = \frac{Kc}{\gamma}\cos(k_u z). \tag{10.18}$$

This is identical with Eq. (2.9) in Chap. 2.

Since the canonical momenta P_x, P_y are zero all along the undulator, the Hamiltonian depends only on z and P_z:

$$H = H(z, P_z) = c\left[\frac{e^2B_0^2}{k_u^2}\cos^2(k_u z) + P_z^2 + m_e^2c^2\right]^{1/2}. \quad (10.19)$$

The Hamilton equations are evaluated using the method of Eq. (10.13):

$$\dot{z} = \frac{\partial H}{\partial P_z} = \frac{P_z}{\gamma m_e}, \quad \dot{P}_z = -\frac{\partial H}{\partial z} = \frac{e^2B_0^2}{2\gamma m_e k_u}\sin(2k_u z).$$

Using $z(t) \approx \bar{v}_z t$ the second equation can be integrated over time. Inserting the result into the first one we obtain

$$v_z(t) \approx \bar{v}_z - \frac{e^2B_0^2}{4\gamma^2 m_e^2 k_u^2 c}\cos(2k_u\bar{v}_z t),$$

where the first term comes from the integration constant. By the same consideration as in Chap. 2 we find that the constant velocity term is identical with the average longitudinal speed (2.11). If we define $\omega_u = \bar{v}_z k_u$ the above expression is identical with Eq. (2.10). The particle trajectory is, choosing $x(0) = y(0) = 0$,

$$x(t) = \frac{cK}{\gamma\omega_u}\sin(\omega_u t), \quad y(t) = 0, \quad z(t) = \bar{v}_z t - \frac{cK^2}{8\gamma^2\omega_u}\sin(2\omega_u t), \quad (10.20)$$

in agreement with Eq. (2.12).

10.1.3 Electron Motion in a Helical Undulator

Generalizing Eq. (10.16) we choose a vector potential

$$\boldsymbol{A} = \frac{B_0}{k_u}\left[\cos(k_u z)\boldsymbol{e}_x + \sin(k_u z)\boldsymbol{e}_y\right] \quad (10.21)$$

which corresponds to the helical undulator field

$$\boldsymbol{B} = -B_0\left[\cos(k_u z)\boldsymbol{e}_x + \sin(k_u z)\boldsymbol{e}_y\right]. \quad (10.22)$$

This field has the sense of rotation of a right-handed screw. Note that it is a simplified field, not in accordance with the Maxwell equation $\nabla \times \boldsymbol{B} = 0$, but it is a good approximation for the field of a real helical undulator for small deviations from the axis. Again the Hamiltonian does not depend on x and y with the consequence that the corresponding canonical momenta are constants of motion and can be set equal to zero by choosing appropriate initial conditions. Hence we have $P_x \equiv 0$, $P_y \equiv 0$, yielding the important equations:

$$\gamma m_e v_x(z) = eA_x(z) = \frac{eB_0}{k_u}\cos(k_u z),$$
$$\gamma m_e v_y(z) = eA_y(z) = \frac{eB_0}{k_u}\sin(k_u z),$$

from which follows

$$v_x = \frac{cK}{\gamma}\cos(k_u z), \quad v_y = \frac{cK}{\gamma}\sin(k_u z). \tag{10.23}$$

Like in the planar undulator the Hamiltonian depends only on z and P_z

$$H = H(z, P_z) = c\left[\frac{e^2 B_0^2}{k_u^2}(\cos^2(k_u z) + \sin^2(k_u z)) + P_z^2 + m_e^2 c^2\right]^{1/2},$$

but now even the z dependence drops out:

$$H = H(P_z) = c\left[\frac{e^2 B_0^2}{k_u^2} + P_z^2 + m_e^2 c^2\right]^{1/2}. \tag{10.24}$$

From this expression follows immediately

$$\dot{P}_z = -\frac{\partial H}{\partial z} = 0 \Rightarrow P_z = \gamma m_e v_z = \text{const.}, \tag{10.25}$$

so the speed in z direction is a constant. From (10.23) we find

$$v_x^2 + v_y^2 = \frac{K^2 c^2}{\gamma^2}$$

and

$$v_z = \sqrt{v^2 - v_x^2 - v_y^2} = c\sqrt{\beta^2 - K^2/\gamma^2} \approx v_0 \equiv c\left(1 - \frac{1}{2\gamma^2}(1 + K^2)\right). \tag{10.26}$$

This differs from the planar undulator case in two aspects: the factor $(1+K^2)$ appears instead of $(1+K^2/2)$, and longitudinal oscillations do not exist. The absence of these oscillations is the reason why helical undulator radiation has no higher harmonics (this applies for radiation in forward direction).

The particle trajectory is found by integration of Eq. (10.23)

$$x(t) = \frac{eB_0}{\gamma m_e k_u^2 v_0} \sin(\omega_u t), \quad y(t) = -\frac{eB_0}{\gamma m_e k_u^2 v_0} \cos(\omega_u t), \quad z(t) = v_0 t. \quad (10.27)$$

It is a right-handed helix with the radius

$$r_{\text{hel}} = \frac{e\,B_0}{\gamma m_e k_u^2 v_0} \approx \frac{K}{\gamma k_u}. \quad (10.28)$$

10.1.4 Energy Exchange Between Electron and Light Wave

10.1.4.1 Planar Undulator

We want to demonstrate that Eq. (3.4), describing the energy transfer between electron and light wave, can also be obtained in the Hamiltonian formalism. Like in Chap. 3 the radiation field is described by a plane wave with linear horizontal polarization

$$E_x(z,t) = E_0 \cos(k_\ell z - \omega_\ell t + \psi_0).$$

The field can be derived from the vector potential

$$\boldsymbol{A}^\ell(z,t) = \frac{E_0}{\omega_\ell} \sin(k_\ell z - \omega_\ell t + \psi_0)\, \boldsymbol{e}_x \quad (10.29)$$

which satisfies the homogeneous wave equation. The total vector potential is the sum of the vector potentials of the undulator (10.16) and the light wave (10.29):

$$\boldsymbol{A}(z,t) = \boldsymbol{A}^u(z) + \boldsymbol{A}^\ell(z,t).$$

It has only an x component but is a function of both z and t. The Hamiltonian of an electron passing a planar undulator in the presence of a horizontally polarized light wave is

$$H(\boldsymbol{r}, \boldsymbol{P}, t) = c\left[\left(P_x + e\,A_x^u(z) + e\,A_x^\ell(z,t)\right)^2 + P_y^2 + P_z^2 + m_e^2\right]^{1/2}. \quad (10.30)$$

The important new feature of this Hamiltonian is its explicit time dependence, brought in by the vector potential of the radiation field. The electron energy is therefore not conserved, which we know already, because there is energy exchange with the radiation field. To find the rate of change dW/dt of the electron energy we compute the partial derivative of H with respect to time using Eq. (10.14). The rate of change of the electron energy is

$$\frac{dW}{dt} = \frac{\partial H}{\partial t} = -e\,v_x(z)E_x(z,t) = -\frac{ecK}{\gamma}\cos(k_u z)E_0\cos(k_\ell z - \omega_\ell t + \psi_0)$$
$$\equiv -\frac{e\,c\,KE_0}{2\gamma}[\cos\psi + \cos\chi] \tag{10.31}$$

which is identical with Eq. (3.4). Here $\psi = (k_\ell + k_u)z - \omega_\ell t + \psi_0$ is the ponderomotive phase defined in Chap. 3, and $\chi = (k_\ell - k_u)z - \omega_\ell t + \psi_0$ is the rapidly oscillating phase. Equation (10.31) is only valid in the low-gain regime because we have treated the field amplitude E_0 of the light wave as a constant when computing the time derivative of the Hamiltonian.

It is very easy to understand that vertically polarized radiation cannot couple to the electron: an electric field having only a y component yields a vanishing scalar product with the velocity vector $\boldsymbol{v} = (v_x, 0, v_z)$ of the electron.

10.1.4.2 Helical Undulator

The helical undulator field (10.22) has a right-handed screw sense. Hence we expect that the helical undulator radiation will be circularly polarized and make the following Ansatz for the light wave interacting with the electrons

$$\boldsymbol{E}(z,t) = E_0\left[\cos(k_\ell z - \omega_\ell t)\boldsymbol{e}_x - \sin(k_\ell z - \omega_\ell t)\boldsymbol{e}_y\right] \tag{10.32}$$

which derives from the vector potential

$$\boldsymbol{A}^\ell(z,t) = \frac{E_0}{\omega_\ell}\left[\sin(k_\ell z - \omega_\ell t)\boldsymbol{e}_x + \cos(k_\ell z - \omega_\ell t)\boldsymbol{e}_y\right]. \tag{10.33}$$

A possible constant phase shift ψ_0 has been omitted here. In the language of modern particle physics the photons described by (10.32) are called right-handed and have positive helicity (a positive projection of angular momentum onto the direction of motion). However, in the nomenclature of optics, Eq. (10.32) describes a left circularly polarized wave.[1] The vector potential of the undulator magnet is given by Eq. (10.21):

$$\boldsymbol{A}^u(z) = \frac{B_0}{k_u}\left[\cos(k_u z)\boldsymbol{e}_x + \sin(k_u z)\boldsymbol{e}_y\right].$$

We can again use Eq. (10.14) to compute the time derivative of the electron energy. The velocities are taken from Eq. (10.23).

---

[1] In particle physics one looks along the direction of motion of the photon. In that case the electric vector at a fixed spatial position rotates in clockwise direction. In optics the convention is such that the observer is facing into the oncoming wave, and then the field vector at a fixed spatial position rotates counterclockwise, see Ref. [4].

$$\frac{dW}{dt} = \frac{\partial H}{\partial t} = -e\,\boldsymbol{v}\cdot\boldsymbol{E} = -ev_x(z)E_x(z,t) - ev_y(z)E_y(z,t)$$
$$= -\frac{ecKE_0}{\gamma}\left[\cos(k_u z)\cos(k_\ell z - \omega_\ell t) - \sin(k_u z)\sin(k_\ell z - \omega_\ell t)\right].$$

It follows

$$\frac{dW}{dt} = -\frac{ecKE_0}{\gamma}\cos\psi \tag{10.34}$$

with the ponderomotive phase $\psi = (k_\ell + k_u)z - \omega_\ell t$ defined in Chap. 3. Sustained energy transfer from the electron to the light wave is obtained if the ponderomotive phase remains constant during the motion through the undulator, the optimum value being $\psi = 0$. From the condition of a stationary phase

$$\dot{\psi} = 0 \Rightarrow (k_\ell + k_u)v_z - k_\ell c = 0$$

we compute the wavelength of helical undulator radiation, using Eq. (10.26):

$$\lambda_\ell = \frac{\lambda_u}{2\gamma^2}\left(1 + K^2\right). \tag{10.35}$$

What happens if we choose the wrong polarization, i.e. right circular polarization (negative helicity)? The field is

$$\boldsymbol{E}(z,t) = E_0\left[\cos(k_\ell z - \omega_\ell t)\boldsymbol{e}_x + \sin(k_\ell z - \omega_\ell t)\boldsymbol{e}_y\right],$$

and one obtains for the time derivative of the electron energy

$$\frac{dW}{dt} = -\frac{ecKE_0}{\gamma}\left[\cos(k_u z)\cos(k_\ell z - \omega_\ell t) + \sin(k_u z)\sin(k_\ell z - \omega_\ell t)\right]$$
$$= -\frac{ecKE_0}{\gamma}\cos\chi. \tag{10.36}$$

In this case the rapidly varying phase $\chi = (k_\ell - k_u)z - \omega_\ell t$ appears which was defined in Chap. 3. As a consequence, the time derivative of the electron energy averages to zero over half an undulator period which means that no energy transfer happens between electron and light wave. In other words: radiation with the wrong circular polarization will not be emitted. From this observation it is obvious that a helical undulator cannot produce linearly polarized radiation either as this is a superposition of left-hand and right-hand circularly polarized waves with 90° relative phase shift. Furthermore we see that it is possible to seed an FEL equipped with a helical undulator with linearly polarized light: a linearly polarized wave can be decomposed into two counter-rotating circularly polarized waves, but only the wave with the correct sense of rotation will be amplified in the FEL. The output light will of course be circularly polarized.

10.2 Supplements to Chapter 4

10.2.1 Derivation of the Third-Order Equation Using the Vlasov Equation

In a one-dimensional description of particle motion in an accelerator we have to specify two quantities for each particle in a bunch: (1) its longitudinal position ζ inside the bunch, and (2) its relative deviation η from the reference energy. These two quantities, which are conjugate variables in a Hamiltonian description of the FEL, define the so-called *longitudinal phase space*. The ponderomotive phase is related to the longitudinal coordinate of an electron inside the bunch, compare Eq. (3.12):

$$\psi = \frac{2\pi\zeta}{\lambda_\ell} - \frac{\pi}{2}. \tag{10.37}$$

Furthermore, it is customary in accelerator physics to replace the independent variable time t by the path length z along the nominal beam orbit. The equations of the one-dimensional high-gain FEL can therefore be written in terms of the variables (ψ, η, z), where z plays the role of a quasi-time.

The ensemble of particles can be described by a distribution function $F(\psi, \eta, z)$ in the (ψ, η) phase space. The number of electrons in the phase-space volume element $d\psi d\eta$ is

$$dn_e = n_e F(\psi, \eta, z) d\psi\, d\eta, \tag{10.38}$$

where n_e is the particle density in the bunch, the number of particles per unit volume. The distribution function is normalized to unity:

$$\frac{1}{2\pi}\int_0^{2\pi}\left(\int F(\psi, \eta, z)\, d\eta\right) d\psi = 1. \tag{10.39}$$

A microscopic expression for the distribution function is presented in Ref. [5].

From our assumption (4.3) of a periodically modulated charge distribution follows that there will be a periodic term in the distribution function, too. Hence we write in complex notation

$$F(\psi, \eta, z) = \Re\left\{\tilde{F}(\psi, \eta, z)\right\} = F_0(\eta) + \Re\left\{\tilde{f}_1(\eta, z)\cdot e^{i\psi}\right\}. \tag{10.40}$$

The modulation amplitude in Eq. (10.40) must remain small to justify the approximations which will be made in the derivation of the third-order equation: $\left|\tilde{f}_1(\eta, z)\right| \ll |F_0(\eta)|$. An example for a periodically modulated particle distribution function is presented in Fig. 10.2.

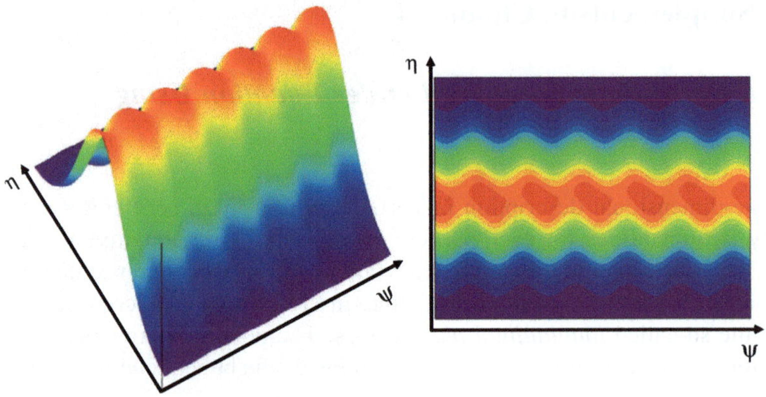

Fig. 10.2 Illustration of a periodically modulated charge density. The particle distribution function is plotted as a function of the ponderomotive phase ψ and the relative energy deviation from resonance $\eta = (W - W_r)/W_r$. Note that in the beginning the periodic modulation is mainly an energy modulation which results from the energy exchange between the electrons and the periodic light wave. The relation between the particle distribution function and the charge density can be found in Eq. (10.43).

For the unperturbed term F_0 we assume a narrow distribution in the relative energy deviation, for example a Gaussian

$$F_0(\eta) = \frac{1}{\sqrt{2\pi}\,\sigma_\eta} \exp\left(-\frac{(\eta - \eta_0)^2}{2\sigma_\eta^2}\right) \quad \text{with} \quad \eta_0 = \frac{W_0 - W_r}{W_r}. \tag{10.41}$$

Note that in general the mean value W_0 of the electron energy may differ slightly from the resonance energy W_r which is defined by the wavelength λ_ℓ of the incident light wave according to Eq. (3.13). The rms spread σ_η is usually small, $\sigma_\eta \le 10^{-4} \ldots 10^{-3}$ for the electron beam in a linac-driven FEL. Hence the relative energy deviation is restricted to a narrow range $|\eta| < \delta$, where $0 < \delta \ll 1$ is chosen such that $F_0(\eta)$ vanishes identically for all $|\eta| \ge \delta$. The Gaussian (10.41) must be truncated at $\eta = \pm\,\delta$ to fulfill this condition.

The integral over the unmodulated part of the distribution function yields

$$\int_{-\delta}^{\delta} F_0(\eta) d\eta = 1. \tag{10.42}$$

The relation between distribution function and charge density or current density is

$$\tilde{\rho}_1(z) = \rho_0 \int_{-\delta}^{\delta} \tilde{f}_1(\eta, z) d\eta, \quad \tilde{j}_1(z) = j_0 \int_{-\delta}^{\delta} \tilde{f}_1(\eta, z) d\eta. \tag{10.43}$$

10.2.1.1 Vlasov Equation

According to the Liouville Theorem of Hamiltonian mechanics, the phase space volume occupied by an ensemble of particles is conserved along the particle trajectory. This leads to a generalized continuity equation which is called the *Vlasov equation* (see for example [6, 7])

$$\frac{dF}{dz} = \frac{\partial F}{\partial z} + \frac{\partial F}{\partial \psi}\frac{d\psi}{dz} + \frac{\partial F}{\partial \eta}\frac{d\eta}{dz} = 0. \tag{10.44}$$

Note that ψ, η and z are real variables. Inserting Eq. (10.40) one obtains

$$\frac{dF}{dz} = \Re\left\{\left(\frac{\partial \tilde{f}_1}{\partial z} + i\tilde{f}_1\frac{d\psi}{dz}\right)e^{i\psi}\right\} + \left(\frac{dF_0}{d\eta} + \Re\left\{\frac{\partial \tilde{f}_1}{\partial \eta}e^{i\psi}\right\}\right)\frac{d\eta}{dz} = 0.$$

The derivative of $\tilde{f}_1$ with respect to η will be neglected here since we assume a small modulation amplitude, $\left|\tilde{f}_1(z,\eta)\right| \ll |F_0(\eta)|$. Using the pendulum equations (4.23) and (4.24) we obtain

$$\Re\left\{\left(\frac{\partial \tilde{f}_1}{\partial z} + i\,2k_u\eta\tilde{f}_1 - \frac{e}{m_e c^2\gamma_r}\frac{dF_0}{d\eta}\left[\frac{\widehat{K}}{2\gamma_r}\tilde{E}_x + \tilde{E}_z\right]\right)e^{i\psi}\right\} = 0. \tag{10.45}$$

This equation holds for any value of ψ. As a consequence, the expression in round brackets must vanish:

$$\frac{\partial \tilde{f}_1}{\partial z} + i\,2k_u\eta\tilde{f}_1 - \frac{e}{m_e c^2\gamma_r}\frac{dF_0}{d\eta}\left[\frac{\widehat{K}}{2\gamma_r}\tilde{E}_x + \tilde{E}_z\right] = 0. \tag{10.46}$$

The proof is straightforward. For $\psi = 0$, the phase factor is $\exp(i\psi) = 1$, hence from Eq. (10.45) follows that the real part of (10.46) is zero. For $\psi = \pi/2$ one has $\exp(i\psi) = i$, so the imaginary part of (10.46) must vanish as well.

The task is now to find a function $\tilde{f}_1$ obeying Eq. (10.46). To this end we replace the longitudinal space charge field by the derivative of the transverse field according to Eq. (4.22) and construct a solution of the following equation

$$\frac{\partial \tilde{f}_1}{\partial z} + i\,2k_u\eta\tilde{f}_1 = \frac{e}{m_e c^2\gamma_r}[\ldots]\frac{dF_0}{d\eta} \tag{10.47}$$

with

$$[\ldots] = \left[\frac{\widehat{K}}{2\gamma_r}\tilde{E}_x + \tilde{E}_z\right] = \left[\frac{\widehat{K}}{2\gamma_r}\tilde{E}_x + i\frac{4\gamma_r c}{\omega_\ell \widehat{K}}\frac{d\tilde{E}_x}{dz}\right]. \tag{10.48}$$

The differential equation (10.47) is of the type

$$y' + i\alpha y(z) = f(z)$$

with the general solution

$$y(z) = \int_0^z f(s)\exp(-i\alpha \cdot (z-s))ds + c_1 \exp(-i\alpha z).$$

Here c_1 is an arbitrary constant. Since the beam is unmodulated at the entrance to the undulator we request $\tilde{f}_1(0) = 0$ and thus $c_1 = 0$, so we obtain

$$\tilde{f}_1(\eta, z) = \frac{e}{m_e c^2 \gamma_r} \int_0^z [\ldots] \frac{dF_0}{d\eta} \exp[-i\, 2k_u \eta \cdot (z-s)]\, ds. \tag{10.49}$$

10.2.1.2 Integro-Differential Equation

The time evolution of the radiation field amplitude is described by the differential equation (4.21) which in combination with Eq. (10.43) yields

$$\frac{d\tilde{E}_x}{dz} = -\frac{\mu_0 c \widehat{K}}{4\gamma_r} \tilde{j}_1 = \frac{\mu_0 c^2 \widehat{K} n_e e}{4\gamma_r} \int_{-\delta}^{\delta} \tilde{f}_1(\eta, z) d\eta. \tag{10.50}$$

Inserting $\tilde{f}_1(\eta, z)$ from (10.49) we get

$$\frac{d\tilde{E}_x}{dz} = \frac{\mu_0 \widehat{K} n_e e^2}{4 m_e \gamma_r^2} \int_0^z \left\{ [\ldots] \int_{-\delta}^{\delta} \frac{dF_0}{d\eta} \exp(-i2k_u \eta \cdot (z-s)) d\eta \right\} ds.$$

Integrating by parts over η, using $F_0(\pm\delta) = 0$, yields an integro-differential equation

$$\frac{d\tilde{E}_x}{dz} = i k_u \frac{\mu_0 \widehat{K} n_e e^2}{2 m_e \gamma_r^2} \int_0^z \left[\frac{\widehat{K}}{2\gamma_r} \tilde{E}_x + i \frac{4\gamma_r c}{\omega_\ell \widehat{K}} \frac{d\tilde{E}_x}{dz} \right] h(z-s) ds \tag{10.51}$$

$$\text{with } h(z-s) = \int_{-\delta}^{\delta} (z-s) \exp[-i2k_u \eta \cdot (z-s)]\, F_0(\eta) d\eta.$$

The two-dimensional problem in the variables (η, z) has now been reduced to a one-dimensional problem in z.

10.2.1.3 Third-Order Equation

The integro-differential equation (10.51) can be simplified if we assume a mono-energetic beam of energy $W = W_0$. Then

$$h(z-s) = (z-s)\,\exp[-i\,2k_u\eta_0\,(z-s)] \quad \text{with } \eta_0 = (W_0 - W_r)/W_r$$

and Eq. (10.51) becomes

$$\frac{d\tilde{E}_x}{dz} = i\,k_u \frac{\mu_0 \widehat{K} n_e e^2}{2m_e\gamma_r^2} \int_0^z [\ldots]\,(z-s)\,\exp[-i\,2k_u\eta_0\,(z-s)]\,ds. \qquad (10.52)$$

In the following we write η instead of η_0, dropping the subscript. Equation (10.52) is of the type

$$y(z) = \int_0^z f(s)(z-s)\exp[-i\alpha\,(z-s)]ds.$$

It is easy to verify that the function $y(z)$ fulfills the differential equation

$$y'' + 2i\alpha y' - \alpha^2 y = f.$$

Applied to Eq. (10.52) this yields

$$\tilde{E}_x''' + 4ik_u\eta\tilde{E}_x'' - 4k_u^2\eta^2\tilde{E}_x' = i\,k_u\frac{\mu_0\widehat{K}n_e e^2}{2m_e\gamma_r^2}\left[\frac{\widehat{K}}{2\gamma_r}\tilde{E}_x + i\frac{4\gamma_r c}{\omega_\ell\widehat{K}}\tilde{E}_x'\right].$$

By ordering terms we finally obtain the well-known third-order differential equation of the high-gain FEL

$$\frac{d^3\tilde{E}_x(z)}{dz^3} + 4ik_u\eta\frac{d^2\tilde{E}_x(z)}{dz^2} + \left(k_p^2 - 4k_u^2\eta^2\right)\frac{d\tilde{E}_x(z)}{dz} - i\Gamma^3\tilde{E}_x(z) = 0. \qquad (10.53)$$

10.2.2 Low-Gain Limit of the High-Gain FEL Theory

In this section we want to demonstrate that the low-gain FEL theory and the Madey theorem can be obtained from the high-gain FEL theory if the undulator magnet is short enough. To this end we start from Eq. (10.52)

$$\frac{d\tilde{E}_x}{dz} = i\,k_u\frac{\mu_0\widehat{K}n_e e^2}{2m_e\gamma_r^2}\int_0^z [\ldots]\,(z-s)\,\exp[-i\,2k_u\eta(z-s)]\,ds$$

with

$$[\ldots] = \left[\frac{\widehat{K}}{2\gamma_r}\tilde{E}_x + i\frac{4\gamma_r c}{\omega_\ell \widehat{K}}\frac{d\tilde{E}_x}{dz} \right].$$

In a short undulator the square bracket can be simplified: the electric field can be considered as roughly constant, $\tilde{E}_x \approx E_0$, and its derivative can be neglected in the square bracket. We then get

$$\begin{aligned} \frac{d\tilde{E}_x}{dz} &= i\,k_u \frac{\mu_0 \widehat{K}^2 n_e e^2 E_0}{4 m_e \gamma_r^3} \int_0^z (z-s)\,\exp\left[-i\,2k_u\eta(z-s)\right] ds \\ &= i\,\Gamma^3 E_0 \int_0^z (z-s)\,\exp\left[-i\,2k_u\eta(z-s)\right] ds. \end{aligned} \tag{10.54}$$

The definite integral yields

$$I(z) \equiv \int_0^z (z-s)\,\exp\left[-i\,2k_u\eta(z-s)\right] ds = \frac{(1+i\,2k_u\eta z)\exp(-i\,2k_u\eta z) - 1}{(2k_u\eta)^2}.$$

The complex field at the end of an undulator of length L_u is

$$\tilde{E}_x(L_u) = E_0 + i\,\Gamma^3 E_0 \int_0^{L_u} I(z) dz \equiv E_0(1 + A(L_u)),$$

where $A(L_u)$ is defined as

$$A(L_u) = -\frac{i\,\Gamma^3}{(2k_u\eta)^2}\left\{ L_u + \frac{i}{k_u\eta} + \left(L_u - \frac{i}{k_u\eta}\right)\exp(-i\,2k_u\eta L_u) \right\}.$$

The absolute square of the field at $z = L_u$ is

$$\left|\tilde{E}_x(L_u)\right|^2 = E_0^2\left(1 + 2\Re[A(L_u)] + (\Re[A(L_u)])^2 + (\Im[A(L_u)])^2\right).$$

The quadratic terms can be neglected in a short undulator, and hence the gain function becomes

$$G = \left| \frac{\tilde{E}_x(L_u)}{E_0} \right|^2 - 1 \approx 2\,\Re[A(L_u)]$$
$$= \frac{2\,\Gamma^3}{(2k_u\eta)^2} \left\{ \frac{1}{k_u\eta} - \frac{\cos(2k_u\eta L_u)}{k_u\eta} - L_u \sin(2k_u\eta L_u) \right\}.$$

Introducing the variable $\xi = k_u\eta L_u$ the gain function can be written as

$$G(\eta) = -\frac{\Gamma^3 L_u^3}{2} \frac{d}{d\xi} \left(\frac{\sin^2 \xi}{\xi^2} \right) \quad \text{with } \xi = \xi(\eta) = k_u L_u \eta = 2\pi N_u \eta. \tag{10.55}$$

This is in perfect agreement with the Madey theorem as stated in Eq. (3.26).

10.2.3 Beam Energy Loss During the FEL Process

In Chap. 4 we have neglected the energy loss of the electron beam which is caused by the continuous energy transfer to the light wave during the FEL gain process. The coupled first-order equations (4.31) are well suited to account for this effect. We take Eq. (4.31b) and average it over the N particles in the FEL bucket:

$$\left\langle \frac{d\eta_n}{dz} \right\rangle = -\frac{e}{m_e c^2 \gamma_r} \Re \left\{ \left(\frac{\widehat{K}\tilde{E}_x}{2\gamma_r} - \frac{i\mu_0 c^2}{\omega_\ell} \cdot \tilde{j}_1 \right) \langle \exp(i\psi_n) \rangle \right\}.$$

From Eq. (4.31c) follows

$$\langle \exp(i\psi_n) \rangle = \frac{1}{N} \sum_{n=1}^{N} \exp(i\psi_n) = \frac{\tilde{j}_1^*}{2j_0}.$$

Inserting this expression into the previous equation yields

$$\frac{d\langle\eta\rangle}{dz} = -\frac{e\widehat{K}}{4\,m_e c^2 \gamma_r^2 j_0} \Re\left\{ \tilde{E}_x(z)\tilde{j}_1^*(z) \right\} + \frac{e\,\mu_0}{2\,m_e \gamma_r j_0 \omega_\ell} \Re\left\{ i\tilde{j}_1(z)\tilde{j}_1^*(z) \right\}.$$

The second term vanishes since $i\tilde{j}_1(z)\tilde{j}_1^*(z) = i\,|\tilde{j}_1(z)|^2$ is purely imaginary. This mathematical result has a deep physical significance: The second term arises from the internal space charge forces among the electrons inside the bunch, and these forces leave the mean electron energy $\langle W \rangle = (1 + \langle\eta\rangle)W_r$ invariant. Hence we arrive at the equation

$$\frac{d\langle\eta\rangle}{dz} = -\frac{e\widehat{K}}{4\,m_e c^2 \gamma_r^2 j_0} \Re\left\{ \tilde{E}_x(z)\tilde{j}_1^*(z) \right\}. \tag{10.56}$$

This equation shows that the differential energy loss of the electron beam grows exponentially along the undulator axis and eventually goes into saturation. The reasoning is as follows. Below saturation, the field $\tilde{E}_x(z)$ grows as $\exp(z/(2L_{g0}))$, and the same is true for $d\tilde{E}_x/dz$. Then Eqs. (10.43) and (10.50) imply that $\tilde{f}_1(\eta, z)$ and $\tilde{j}_1(z)$ also grow as $\exp(z/(2L_{g0}))$. Therefore the differential energy loss grows as $\exp(z/L_{g0})$, just like the FEL power. This is exactly what one would expect intuitively.

When the FEL power goes into saturation, also the term $\Re\left\{\tilde{E}_x(z)\tilde{j}_1^*(z)\right\}$ approaches its maximum value.

10.2.4 Non-periodic First-Order Equations

Our goal is to generalize the first-order equations in such a way that non-periodic processes in the FEL can be handled. Examples are the SASE mechanism where the initial particle distribution is random, FEL seeding with a superposition of harmonic waves of different frequencies, non-uniform charge density profiles in the electron bunches, effects at the head or tail of the electron bunch, or slippage between the electrons and the light wave.

In the periodic model, outlined in Sect. 4.6, the slowly varying amplitudes depend only on the position z in the undulator. Field and current density have been written as

$$\tilde{E}_x(z,t) = \tilde{E}_x(z)\exp[i\,k_\ell z - i\,\omega_\ell t],$$
$$\tilde{j}_z(\psi, z) = j_0 + \tilde{j}_1(z)\,\exp\left[i\,(k_\ell + k_u)z - i\,\omega_\ell t\right].$$

In the non-periodic generalization the current density inside the electron bunch depends in addition on our previously defined internal bunch coordinate

$$\zeta = z - \bar{\beta}\,c\,t = z - \bar{v}_z t$$

and the electric field depends on a corresponding coordinate u inside the FEL pulse

$$u = z - c\,t = \left(1 - \frac{c}{\bar{v}_z}\right) z + \frac{c}{\bar{v}_z}\,\zeta.$$

Hence the Ansatz is made

$$\tilde{E}_x(z,t) = \widehat{E}(z,u)\exp[i\,k_\ell z - i\,\omega_\ell t], \tag{10.57}$$

$$\tilde{j}_z(z,t) = j_0(\zeta) + \widehat{j}_1(z,\zeta)\exp[i\,(k_\ell + k_u)z - i\,\omega_\ell t]. \tag{10.58}$$

The complex field amplitude is now denoted with $\widehat{E}(z,u)$, and the subscript x is dropped for simplicity. These expressions are substituted into the wave equation (4.9).

$$\left[\frac{\partial^2}{\partial z^2} - \frac{1}{c^2}\frac{\partial^2}{\partial t^2}\right]\tilde{E}_x(z,t) = \mu_0 \frac{\partial \tilde{j}_x}{\partial t}. \tag{10.59}$$

Then we obtain

$$\frac{\partial^2 \tilde{E}_x(z,t)}{\partial z^2} = \left[\frac{\partial^2}{\partial z^2} + 2\frac{\partial^2}{\partial z \partial u} + \frac{\partial^2}{\partial u^2} + 2ik_\ell\left(\frac{\partial}{\partial z} + \frac{\partial}{\partial u}\right) - k_\ell^2\right]\widehat{E}(z,u)$$
$$\times \exp[i k_\ell z - i\omega_\ell t],$$
$$\frac{\partial^2 \tilde{E}_x(z,t)}{\partial t^2} = c^2\left[\frac{\partial^2}{\partial u^2} + 2ik_\ell \frac{\partial}{\partial u} - k_\ell^2\right]\widehat{E}(z,u)\exp[i k_\ell z - i\omega_\ell t].$$

In the slowly varying amplitude (SVA) approximation, $\widehat{E}(z,u)$ is assumed to be a smooth and slowly varying function of z and u, so that its second derivatives can be neglected:

$$\left|\frac{\partial^2 \widehat{E}}{\partial z^2}\right| \ll k_\ell \left|\frac{\partial \widehat{E}}{\partial z}\right|, \quad \left|\frac{\partial^2 \widehat{E}}{\partial z \partial u}\right| \ll k_\ell \left|\frac{\partial \widehat{E}}{\partial z}\right|.$$

Then the left-hand side of the wave equation becomes

$$\left[\frac{\partial^2}{\partial z^2} - \frac{1}{c^2}\frac{\partial^2}{\partial t^2}\right]\tilde{E}_x(z,t) = 2i k_\ell \frac{\partial}{\partial z}\widehat{E}(z,u)\, \exp[i k_\ell z - i\omega_\ell t]. \tag{10.60}$$

The current density $\tilde{j}_x$ appearing on the right-hand side of (10.59) is written in the form (compare Eq. (4.15))

$$\tilde{j}_x \approx \frac{v_x}{c}\tilde{j}_z = \left[j_0(\zeta) + \widehat{j}_1(z,\zeta)\exp[i(k_\ell + k_u)z - i\omega_\ell t]\right]\frac{K}{\gamma}\cos(k_u z).$$

The stimulation by the shape term $j_0(\zeta)$ is far away from the FEL resonance, and hence this term can be dropped when computing the time derivative:

$$\frac{\partial \tilde{j}_x}{\partial t} = -\left[\bar{v}_z \frac{\partial \widehat{j}_1}{\partial \zeta} + i\omega_\ell \widehat{j}_1(z,\zeta)\right]\exp[i(k_\ell + k_u)z - i\omega_\ell t]\frac{K}{\gamma}\cos(k_u z).$$

In the SVA approximation

$$\bar{v}_z \left|\frac{\partial \widehat{j}_1}{\partial \zeta}\right| \ll \left|\omega_\ell \widehat{j}_1(z,\zeta)\right|,$$

so the wave equation (10.59) reduces to

$$\frac{\partial}{\partial z}\widehat{E}(z,u) = -\frac{\mu_0 c K}{2\gamma}\widehat{j}_1(z,\zeta)\exp(ik_u z)\cos(k_u z).$$

The average value of $\exp(i\,k_u z)\cos(k_u z)$ over one undulator period is $1/2$. Replacing K by $\widehat{K}$ to take the longitudinal oscillation into account we finally obtain the generalization of Eq. (4.31d):

$$\frac{\partial}{\partial z}\widehat{E}(z,u) = -\frac{\mu_0 c \widehat{K}}{4\gamma_r}\widehat{j}_1(z,\zeta). \tag{10.61}$$

The microscopic expression for the current density is

$$j_z = \frac{Q_b\,\bar{v}_z}{N_t A_b}\sum_{n=1}^{N_t}\delta(\zeta - \zeta_n(z)),$$

where $Q_b = -N_t\,e$ is the charge of the bunch, A_b is the cross section, and N_t is the total number of electrons in the bunch. Since the present FEL codes work with a far smaller number N of test particles, j_z must be approximated by smooth functions $j_0(\zeta)$ and $\widehat{j}_1(z,\zeta)$ that vary slowly in the z coordinate (compared to the undulator period λ_u) and in the ζ coordinate (compared to the FEL wavelength λ_ℓ). Indeed, the conversion of the particle positions ζ_n to smooth functions is a delicate and sensitive part of FEL simulations using macroparticle descriptions. Pseudo-random distributions can be used yielding the same statistical fluctuations that are present in the real electron beam.

In most FEL codes locally periodic conditions are assumed. The bunch is subdivided into slices of length λ_ℓ which are similar the FEL buckets. Within each slice periodic conditions are assumed. The local amplitude of the first harmonic is written as

$$\widehat{j}_1(z,c_m) \approx j_0(c_m)\frac{2}{N_m}\sum_{n\in I_m}\exp(-i\,k_\ell\zeta_n), \tag{10.62}$$

where $c_m = m\,\lambda_\ell$ is the center of slice m, N_m is the number of particles in that slice, and I_m is the index range.

We know from Sect. 5.6 that not all particles stay in their bucket during the FEL gain process but many of them move into the next bucket. In a simulation program this is accounted for by replacing particles leaving a slice on one side by corresponding particles entering the slice from the other side. Further details are beyond the scope of this book. The non-periodic form of the coupled equations is

$$\frac{d\psi_n}{dz} = 2k_u\eta_n, \quad n = 1, \ldots, N, \tag{10.63a}$$

$$\frac{d\eta_n}{dz} = \frac{-e}{m_e c^2 \gamma_r} \Re\left\{ \left(\frac{\widehat{K}\,\widehat{E}(z, u_n)}{2\gamma_r} - \frac{i\mu_0 c^2 \widehat{j}_1(z, \zeta_n)}{\omega_\ell} \right) e^{i\psi_n} \right\}, \tag{10.63b}$$

$$\widehat{j}_1(z, c_m) = j_0(c_m)\, \frac{2}{N_m} \sum_{n \in I_m} \exp(-i\, k_\ell \zeta_n), \quad c_m = m\,\lambda_\ell, \tag{10.63c}$$

$$\frac{\partial \widehat{E}(z, u)}{\partial z} = -\frac{\mu_0 c \widehat{K}}{4\gamma_r} \widehat{j}_1(z, \zeta). \tag{10.63d}$$

Here $\zeta_n = (\psi_n + \pi/2)\,\lambda_\ell/(2\pi)$ and $u_n \approx \zeta_n - (1 - \bar{\beta})z$. In these equations, N is the total number of test particles per bunch and not the number of test particles per slice as in the periodic case.

10.3 Gaussian Modes of Laser Beams

Gaussian beam modes are characteristic of the output of lasers. They are also well suited to describe the beam in a free-electron laser. In this section we follow essentially the treatments in Refs. [8] and [9].

10.3.1 Fundamental Gaussian Mode

Laser beams are coherent electromagnetic radiation and obey the Maxwell equations. In a medium with refractive index n the electric field vector satisfies the wave equation

$$\nabla^2 E - \frac{n^2}{c^2}\frac{\partial^2 E}{\partial t^2} = 0, \tag{10.64}$$

where $E(x, y, z, t)$ can be any Cartesian component of $\boldsymbol{E}$. We restrict ourselves to homogeneous media (n =const) or vacuum ($n = 1$). In the laser literature often quadratic index media are considered, see e.g. [8, 9]. We look for a cylindrically symmetric solution of the wave equation which depends only on $r = \sqrt{x^2 + y^2}$, z, and t, but not on the azimuthal angle. Writing

$$E(x, y, z, t) = f(r, z)\exp(i\,kz - i\,\omega t) \quad \text{with } k = n\omega/c$$

we obtain the equation

$$\frac{\partial^2 f}{\partial r^2} + \frac{1}{r}\frac{\partial f}{\partial r} + 2\,i\,k\frac{\partial f}{\partial z} = 0. \tag{10.65}$$

In practice the field amplitude varies slowly with z, hence the second derivative $\partial^2 f/\partial z^2$ can be neglected in comparison with $2\,i\,k\,\partial f/\partial z$ (SVA approximation). The simplest solution has no zeros in radial direction. It is convention to write it in the form

$$f(r, z) = \exp\left[iP(z) + \frac{ikr^2}{2q(z)}\right]$$

with two complex functions $P(z)$ and $q(z)$. Substituting this into Eq. (10.65) we obtain

$$-\left(\frac{k}{q}\right)^2 r^2 + 2i\left(\frac{k}{q}\right) - 2kP' - k^2 r^2\left(\frac{1}{q}\right)' = 0,$$

where the prime indicates differentiation with respect to z. This equation holds for all r, so the coefficients of different powers of r must vanish:

$$\left(\frac{1}{q}\right)^2 + \left(\frac{1}{q}\right)' = 0, \quad P' = \frac{i}{q}. \tag{10.66}$$

The solutions are

$$q(z) = z + q_0\,, \quad P(z) = i\,\ln\left(1 + \frac{z}{q_0}\right) \tag{10.67}$$

with a complex constant q_0. The amplitude function becomes

$$f(r,z) = \exp\left[-\ln\left(1 + \frac{z}{q_0}\right) + \frac{i\,k\,r^2}{2(q_0 + z)}\right]. \tag{10.68}$$

The amplitude must vanish in the limit $r \to \infty$ for any value of z. At $z = 0$ this is realized if the constant q_0 is a negative imaginary number. It is expressed in the form

$$q_0 = -i\,z_R \quad \text{with } z_R = \frac{k\,w_0^2}{2} = \frac{\pi w_0^2}{\lambda/n}. \tag{10.69}$$

Here z_R is the so-called *Rayleigh length*, a characteristic length for diffraction, and $\lambda = 2\pi c/\omega$ is the wavelength in vacuum.

The laser beam is assumed to have a waist of width w_0 at $z = 0$. Furthermore the following quantities are defined

$$w(z) = w_0\sqrt{1 + (z/z_R)^2}, \quad R(z) = z\left(1 + (z_R/z)^2\right). \tag{10.70}$$

We will see below that the beam width at position z is described by $w(z)$ while $R(z)$ is the radius of curvature of the wave front. Then

$$\frac{1}{q(z)} = \frac{1}{R(z)} + i\,\frac{\lambda/n}{\pi w^2(z)}. \tag{10.71}$$

With these quantities the first exponential term in the amplitude function (10.68) can be written in the form

$$\exp\left[-\ln\left(1 + i\,z/z_R\right)\right] = \frac{w_0}{w(z)}\exp(-i\,\chi(z))$$

with the so-called *Gouy phase*

$$\chi(z) = \arctan(z/z_R). \tag{10.72}$$

We have used $\ln(a + i\,b) = \ln(\sqrt{a^2 + b^2}) + i\,\arctan(b/a)$. The second exponential term in Eq. (10.68) becomes after separating real and imaginary parts

$$\exp\left[\frac{i\,k\,r^2}{2q(z)}\right] = \exp\left[-\frac{r^2}{w^2(z)}\right]\exp\left[\frac{i\,k\,r^2}{2R(z)}\right].$$

The electric field of a horizontally polarized laser wave is

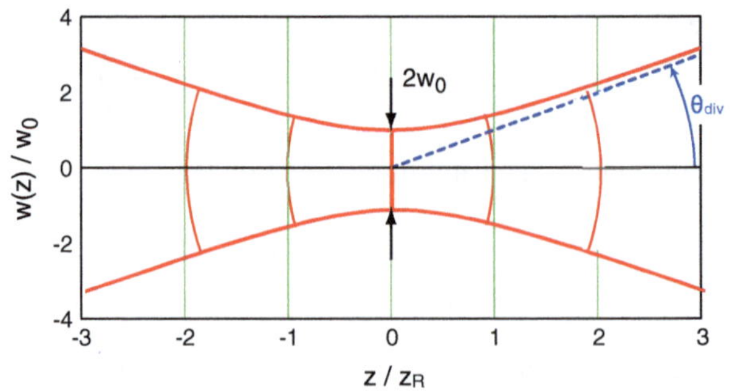

Fig. 10.3 The beam envelope $\pm w(z)/w_0$ in the vicinity of a waist. The *dotted line* indicates the divergence angle $\theta_{\rm div}$. The *spherical* wavefronts at $z = \pm z_R$ and $z = \pm 2\,z_R$ are sketched.

$$E_x(r,z,t) = E_0\,\frac{w_0}{w(z)}\,\exp\left[-\frac{r^2}{w^2(z)}\right]\,\exp\left[ikz - i\chi(z) - i\,\omega t\right]\,\exp\left[\frac{ikr^2}{2R(z)}\right]$$
$$= E_0\,\exp\left(\frac{-(r/w_0)^2}{1+i\,z/z_R}\right)\,\frac{\exp(ikz - i\omega t)}{1+i\,z/z_R}. \tag{10.73}$$

There are many differences to a plane wave:

- The field amplitude drops to $1/e$ of its peak value at a radius $r = w(z)$. The smallest width w_0 is obtained at the waist at $z = 0$.
- The phase velocity is different from that of a plane wave, which is $v_{\rm ph} = \omega/k = c/n$. The presence of the Gouy phase changes the phase velocity of the Gaussian laser wave near a beam waist.
- The phase factor $\exp\left[i\,k\,r^2/(2R(z))\right]$ implies that the wave fronts are curved and of nearly spherical shape.

The beam is sketched in Fig. 10.3. The radius of curvature becomes infinite at the waste position $z = 0$ and assumes its minimum value of $R_{\rm min} = 2z_R$ at $z = z_R$. Due to the divergence of the beam the light rays are not exactly parallel to the z axis. As a consequence the electric field has a small z component. This follows also from the first Maxwell equation $\nabla \cdot \boldsymbol{E} = 0$: the radial derivative is non-zero, hence a longitudinal field component E_z must be present.

The divergence angle of the beam emerging from the waist is

$$\theta_{\rm div} \approx \tan\theta_{\rm div} = \frac{\lambda}{n\,\pi\,w_0}. \tag{10.74}$$

The product of beam width and divergence is for a beam in vacuum

$$w_0 \cdot \theta_{\rm div} = \frac{\lambda}{\pi}. \tag{10.75}$$

The intensity of the laser beam is quadratic in the electric field and thus proportional to $\exp\left[-2r^2/w^2(z)\right]$. In terms of Gaussian variances one gets at $z = 0$

$$\sigma_x = w_0/2\,, \quad \sigma_\theta = \theta_{\text{div}}/2.$$

From (10.75) follows then the fundamental relation between laser beam width and beam divergence

$$\sigma_x \cdot \sigma_\theta = \frac{\lambda}{4\pi}. \tag{10.76}$$

In accelerator terminology this product is called the beam emittance.

10.3.2 Higher-Order Gaussian Beam Modes

The fundamental Gaussian mode depends only the distance r from the axis and the longitudinal coordinate z. If we do not impose cylindrical symmetry the wave equation (10.64) has solutions of the form [9]

$$E_x(x,y,z) = E_0\,\frac{w_0}{w(z)}\,H_m\left(\frac{\sqrt{2}\,x}{w(z)}\right)H_n\left(\frac{\sqrt{2}y}{w(z)}\right)\exp\left[-\frac{x^2+y^2}{w^2(z)}\right]$$
$$\times\,\exp\left[ikz - i(m+n+1)\chi(z)\right]\,\exp\left[\frac{ik(x^2+y^2)}{2R(z)}\right]. \tag{10.77}$$

The H_m are the Hermite polynomials. The fundamental mode is the special case $m = n = 0$. The TEM_{00} mode is identical with the fundamental Gaussian mode discussed in Sect. 10.3.1. This mode has its highest intensity on the axis. In an FEL there is optimum overlap between the TEM_{00} mode and the electron beam, and for this reason this mode will be strongly amplified. The higher modes with odd indices have vanishing intensity on the axis and can generally be neglected in the high-gain FEL while the modes with even indices have a finite size on the beam axis. In the TEM_{10} mode the electric field E_x changes sign when going from positive to negative x. This is because $H_1(x)$ is an odd function. Therefore this mode cannot couple to an electron beam with a charge distribution that is symmetric in x. The electric field pattern of the TEM_{00} and TEM_{10} Hermite–Gaussian laser modes is shown in Fig. 10.4. In the TEM_{20} and TEM_{22} modes (see Fig. 10.5) the electric field E_x is an even function of x. These modes couple to a symmetric electron beam.

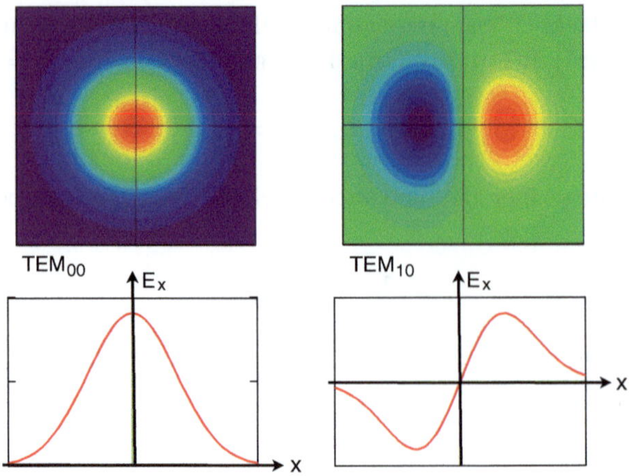

Fig. 10.4 *Top*: Electric field distribution of the TEM_{00} and TEM_{10} Hermite–Gaussian laser modes in a plane transverse to the axis of propagation. *Bottom*: The field $E_x(x, 0, 0)$ of the TEM_{00} and TEM_{10} modes as a function of the horizontal coordinate x at $y = z = 0$.

Fig. 10.5 Electric field distribution of the TEM_{20} and TEM_{22} Hermite–Gaussian laser modes.

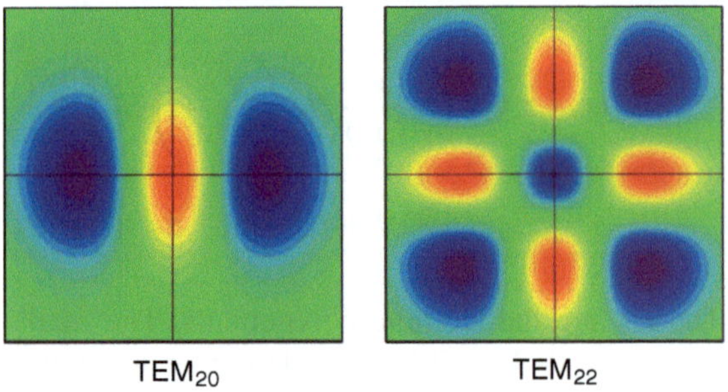

10.4 Eigenmode Approach of 3D FEL

We have seen in Chap. 4 that the third-order equation for the slowly varying field amplitude $\tilde{E}_x(z)$ is solved by an exponential function $\exp(\alpha z)$. For a mono-energetic beam which is on resonance ($\eta = (W - W_r)/W_r = 0$) and for negligible space charge ($k_p = 0$), the three eigenvalues are solutions of the simple cubic equation

$$\alpha_j^3 = i\,\Gamma^3 .$$

In this section we consider more general cases. For a beam energy different from the resonance energy ($W \neq W_r$, $\eta \neq 0$) and with space charge present ($k_p \neq 0$) the third-order equation is still applicable, and for reasonable values of these parameters there exists one eigenvalue α_1 with positive real part leading to exponential growth. However, the third-order equation is not applicable if beam energy spread is present ($\sigma_\eta > 0$). In this case the linear integro-differential equation (10.51) can be used which will usually possess an eigenvalue with $\Re(\alpha_1) > 0$, provided σ_η is small enough to preserve the main features of the solutions. The eigenvalue leading to exponential growth will be determined in the following sections. First the general procedure is described, resulting in Eq. (10.88), and then it is applied to the one-dimensional FEL and a simplified version of the three-dimensional FEL. We will show that in the 1D case the two dimensionless quantities $\sigma_\eta/\rho_{\mathrm{FEL}}$ and k_p/Γ have a significant influence on the FEL gain. In the 3D case the relative size of the product $L_{g0}\lambda_\ell$ and the beam cross section A_b play a role in addition.

10.4.1 General Procedure

In this section we generalize the method for determining the eigenvalues α_j to the 3D case. To this end the full three-dimensional wave equation for the electric field of a horizontally polarized light wave has to be considered

$$\left[\nabla^2 - \frac{1}{c^2}\frac{\partial^2}{\partial t^2}\right] E_x(x, y, z, t) = \mu_0 \frac{\partial j_x}{\partial t} + \frac{1}{\varepsilon_0}\frac{\partial \rho}{\partial x} = \frac{1}{\varepsilon_0}\left(\frac{1}{c^2}\frac{\partial j_x}{\partial t} + \frac{\partial \rho}{\partial x}\right).$$

The second term on the right-hand side can be neglected for two reasons:

(1) Its magnitude is usually small compared to that of the first term, for example less than 5 % at FLASH.
(2) More importantly, this term has a phase slippage $\exp(ik_u z)$ against the FEL wave and its contribution to the growth rate of the field averages out over one undulator period. This can be seen from the equation

$$\frac{\partial \rho}{\partial x} = \frac{\partial \tilde{\rho}_1}{\partial x}\exp(i\psi) = \frac{\partial \tilde{\rho}_1}{\partial x}\exp(ik_\ell z - i\omega_\ell t)\exp(ik_u z).$$

The phase of $\partial\tilde{\rho}_1/\partial x$ varies slowly (SVA approximation) and can be taken as constant over one undulator period.

In the present eigenmode approach we neglect betatron oscillations. The transverse offset $\boldsymbol{r}_\perp = (x, y)$ appears as an independent coordinate in the distribution function but the derivatives with respect to x and y will be dropped in the Vlasov equation. For the transverse electric field the Ansatz is made

$$\tilde{E}_x(\boldsymbol{r}_\perp, z, t) = \tilde{E}_x(\boldsymbol{r}_\perp, z)\exp[i(k_\ell z - \omega_\ell t)] \quad \text{with } \boldsymbol{r}_\perp = (x, y)\,. \tag{10.78}$$

Applying the slowly varying amplitude approximation and expressing j_x in terms of $\tilde{j}_1$, this equation assumes the form

$$\left[\nabla_\perp^2 + 2\,i\,k_\ell\frac{\partial}{\partial z}\right]\tilde{E}_x = -\frac{i\,k_\ell\mu_0 c\,\widehat{K}}{2\gamma_r}\tilde{j}_1(\boldsymbol{r}_\perp, z) \quad \text{with } \nabla_\perp^2 = \frac{\partial^2}{\partial x^2} + \frac{\partial^2}{\partial y^2}. \tag{10.79}$$

This is the three-dimensional generalization of Eq. (4.21).

In the 3D case the distribution function acquires a dependency on the transverse coordinates. To keep the relation to the 1D model, an equivalent beam cross section A_b is chosen which is used for normalization purposes but has no influence on absolute (unnormalized) quantities. We set

$$A_b = \frac{I_0}{\max(|j_0|)},$$

where I_0 is the magnitude of the dc beam current and $\max(|j_0|)$ the maximum value of the dc current density. For a round beam with constant charge density we get $A_b = \pi r_b^2$, and for a Gaussian beam with $\sigma_x = \sigma_y = \sigma_r$ one gets $A_b = 2\pi\sigma_r^2$. The dc beam current is $I_0 = n_e A_b\, e\, c$.

Generalizing Eq. (10.40) the particle distribution function is written as

$$F(\boldsymbol{r}_\perp, \psi, \eta, z) = F_0(\boldsymbol{r}_\perp, \eta) + \Re\left\{\tilde{f}_1(\boldsymbol{r}_\perp, \eta, z)\cdot e^{i\psi}\right\}. \tag{10.80}$$

The dc current density and the modulated current density are according to Eq. (10.43)

$$j_0(\boldsymbol{r}_\perp) = -I_0\int_{-\delta}^{\delta} F_0(\boldsymbol{r}_\perp, \eta)\,d\eta, \quad \tilde{j}_1(\boldsymbol{r}_\perp, z) = -I_0\int_{-\delta}^{\delta}\tilde{f}_1(\boldsymbol{r}_\perp, \eta, z)\,d\eta. \tag{10.81}$$

Remember that I_0 is the absolute magnitude of the dc current and thus positive, while j_0 is negative since the electron charge is $q_e = -e$. The modulated distribution function obeys an equation similar to Eq. (10.47)

$$(\partial_z + i2k_u\eta)\tilde{f}_1(\boldsymbol{r}_\perp, \eta, z) = \frac{e}{m_e c^2\gamma_r}\left[\frac{\widehat{K}}{2\gamma_r}\tilde{E}_x(\boldsymbol{r}_\perp, z) + \tilde{E}_z(\boldsymbol{r}_\perp, z)\right]\frac{dF_0}{d\eta}. \tag{10.82}$$

The longitudinal space charge field $\tilde{E}_z$ can be computed from the modulated charge density using the Maxwell equation $\nabla \cdot \boldsymbol{E} = \rho/\varepsilon_0$. Neglecting the derivatives with respect to the transverse coordinates x and y, and using the SVA approximation one finds as a generalization of Eq. (4.18)

$$\tilde{E}_z(\boldsymbol{r}_\perp, z) = -\frac{i\mu_0 c}{k_\ell}\tilde{j}_1(\boldsymbol{r}_\perp, z) = \frac{i\mu_0 c\, I_0}{k_\ell}\int \tilde{f}_1(\boldsymbol{r}_\perp, \eta, z)\, d\eta. \tag{10.83}$$

It is the spirit of the linear eigenmode analysis to assume that $\tilde{E}_x$ and $\tilde{f}_1$ depend exponentially on z:

$$\tilde{E}_x(\boldsymbol{r}_\perp, z) = A(\boldsymbol{r}_\perp)e^{\alpha z}, \quad \tilde{f}_1(\boldsymbol{r}_\perp, \eta, z) = B(\boldsymbol{r}_\perp, \eta)e^{\alpha z}. \tag{10.84}$$

Substitution in Eqs. (10.79) and (10.82) yields the equations

$$\left[\nabla_\perp^2 + 2\,i\,k_\ell\alpha\right]A(\boldsymbol{r}_\perp) = \frac{i\,k_\ell\mu_0 c\, I_0\widehat{K}}{2\gamma_r}\int B(\boldsymbol{r}_\perp, \eta)\, d\eta, \tag{10.85}$$

$$[\alpha + 2\,i\,k_u\eta]\, B(\boldsymbol{r}_\perp, \eta) = \frac{e}{m_e c^2\gamma_r}\left[\frac{\widehat{K}}{2\gamma_r}A(\boldsymbol{r}_\perp) + \frac{i\mu_0 c\, I_0}{k_\ell}\int B(\boldsymbol{r}_\perp, \eta)\, d\eta\right]\frac{dF_0}{d\eta}. \tag{10.86}$$

The quantities to be determined are the eigenvalues α_j and the coefficient functions $A(\boldsymbol{r}_\perp)$ and $B(\boldsymbol{r}_\perp, \eta)$. Note that the integral

$$\int B(\boldsymbol{r}_\perp, \eta)\, d\eta = \int_{-\delta}^{\delta} B(\boldsymbol{r}_\perp, \eta)\, d\eta$$

depends only on the transverse coordinates $\boldsymbol{r}_\perp$ but not on the fractional energy deviation η. This allows us to divide Eq. (10.86) by $[\alpha + 2\,i\,k_u\eta]$ to obtain $B(\boldsymbol{r}_\perp, \eta)$, and then to integrate over η.

$$\int_{-\delta}^{\delta} B(\boldsymbol{r}_\perp, \eta)\, d\eta = \frac{e}{m_e c^2\gamma_r}\left[\frac{\widehat{K}}{2\gamma_r}A(\boldsymbol{r}_\perp) + \frac{i\mu_0 c\, I_0}{k_\ell}\int_{-\delta}^{\delta} B(\boldsymbol{r}_\perp, \eta)\, d\eta\right]$$
$$\times \int_{-\delta}^{\delta}\frac{1}{[\alpha + 2\,i\,k_u\eta]}\frac{dF_0}{d\eta}\, d\eta.$$

(For the definition of the integration limits $\pm\delta$ see Eq. (10.42)).

Now we make the assumption that there is no correlation between energy spread and transverse offset in the unmodulated part F_0 of the distribution function so that F_0 can be factorized

$$F_0(\boldsymbol{r}_\perp, \eta) = G(\boldsymbol{r}_\perp)H(\eta). \tag{10.87}$$

Integrating by parts and using $H(\pm\delta) = 0$ we obtain

$$\begin{aligned}\int B(\boldsymbol{r}_\perp, \eta)d\eta &= \frac{i\,2k_u\,e}{m_e c^2 \gamma_r}\left[\frac{\widehat{K}}{2\gamma_r}A(\boldsymbol{r}_\perp) + \frac{i\mu_0 c I_0}{k_\ell}\int B(\boldsymbol{r}_\perp, \eta)d\eta\right]G(\boldsymbol{r}_\perp)\,C(\alpha)\\ &= \left[\frac{i\,k_u e\,\widehat{K}}{m_e c^2 \gamma_r^2}A(\boldsymbol{r}_\perp) - k_p^2 A_b \int B(\boldsymbol{r}_\perp, \eta)d\eta\right]G(\boldsymbol{r}_\perp)C(\alpha)\end{aligned}$$

with the abbreviations

$$C(\alpha) = \int \frac{H(\eta)}{(\alpha + i2k_u\eta)^2}d\eta, \quad k_p^2 = \frac{2k_u\mu_0 e}{m_e c\gamma_r k_\ell}\frac{I_0}{A_b}.$$

Then we can express $\int B(\boldsymbol{r}_\perp, \eta)d\eta$ in terms of $A(\boldsymbol{r}_\perp)$:

$$\int B(\boldsymbol{r}_\perp, \eta)d\eta = \frac{ik_u e\widehat{K}}{m_e c^2 \gamma_r^2}\,\frac{CA_bG}{1 + k_p^2 CA_bG}A(\boldsymbol{r}_\perp).$$

When this is substituted in Eq. (10.85) we get an implicit equation for the determination of the eigenvalues α_j:

$$\left[\nabla_\perp^2 + 2ik_\ell\alpha\right]A(\boldsymbol{r}_\perp) = (2ik_\ell)\cdot i\Gamma^3\frac{C(\alpha)A_bG(\boldsymbol{r}_\perp)}{1 + k_p^2 C(\alpha)A_bG(\boldsymbol{r}_\perp)}A(\boldsymbol{r}_\perp). \tag{10.88}$$

Note that the gain parameter Γ and the space charge parameter k_p are defined in the same way as in the one-dimensional theory, see Eq. (4.44).

10.4.2 One-Dimensional Case

In the 1D case the function $A(\boldsymbol{r}_\perp) \equiv A$ is a constant and the operator $\nabla_\perp^2$ can be omitted from Eq. (10.88). Moreover, we can replace $H(\eta)$ by $F_0(\eta)$ and $A_bG(\boldsymbol{r}_\perp)$ by 1. Dividing Eq. (10.88) by A and ordering terms we obtain the following eigenvalue equation

$$\alpha = \left[i\Gamma^3 - k_p^2\alpha\right]\int_{-\delta}^{\delta} \frac{F_0(\eta)}{(\alpha + i2k_u\eta)^2}d\eta. \tag{10.89}$$

Note that this eigenvalue equation is valid for an arbitrary initial energy distribution $F_0(\eta)$ in the beam. An alternative derivation of this result is possible by inserting the Ansatz $\tilde{E}_x = A\exp(\alpha z)$ into Eq. (10.51).

The power growth in the exponential regime is determined by the real part of α_1:

$$P(z) \propto \exp(2\Re(\alpha_1)z).$$

The growth rate depends on the mean relative energy offset

$$\langle\eta\rangle = \int_{-\delta}^{\delta} \eta F_0(\eta)d\eta.$$

The maximum value of the function $2\Re(\alpha_1(\eta))$ yields the fastest power rise and is thus directly related to the power gain length. It is therefore meaningful to define the power gain length by the equation

$$L_g = \frac{1}{\max\{2\Re(\alpha_1(\eta))\}}. \tag{10.90}$$

We can convince ourselves that this expression reduces to the 1D power gain length L_{g0} for a beam with $\eta = 0$ and $k_p = 0$. In that case we have

$$\alpha_1 = (i + \sqrt{3})\Gamma/2 \Rightarrow \frac{1}{2\Re(\alpha_1)} = \frac{1}{\sqrt{3}\Gamma} = L_{g0}.$$

The above discussion shows that it is useful to introduce a normalized growth rate function

$$f_{gr}(\eta) = 2\Re(\alpha_1(\eta))L_{g0}.$$

The maximum value of this function is identical with the ratio L_{g0}/L_g.

For the special case of a mono-energetic beam of energy W, Eq. (10.89) becomes

$$\alpha^3 + i4k_u\eta\alpha^2 + (k_p^2 - 4k_u^2\eta^2)\alpha - i\Gamma^3 = 0. \tag{10.91}$$

Alternatively, this result is obtained by inserting the Ansatz $\tilde{E}_x = A\exp(\alpha z)$ into Eq. (10.53). The three roots α_j can in principle be computed analytically, although this may be cumbersome. Often a numerical computation is easier.

An analytical approach is reasonably straightforward for the case $k_p = 0$. Equation (10.91) can be transformed into a normalized algebraic equation

$$a\,(a + ib)^2 - i = 0 \quad \text{with} \quad a = \frac{\alpha}{\Gamma} \quad \text{and} \quad b = \frac{\eta}{\rho_{\text{FEL}}}. \tag{10.92}$$

The three solutions of this cubic equation are

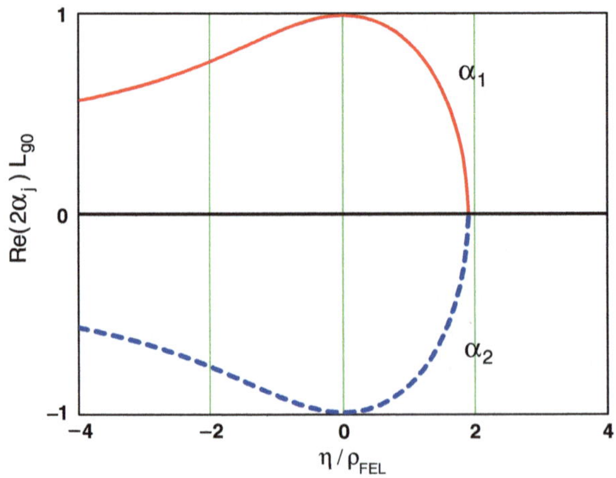

Fig. 10.6 The real part of the first and second eigenvalue, multiplied with $2L_{g0}$, is plotted as a function of η/ρ_{FEL}, the relative energy deviation divided by the FEL parameter. Note that $\Re(\alpha_1)$ (*continuous red curve*) is positive, corresponding to exponential growth of the eigensolution $V_1(z) = \exp(\alpha_1 z)$. Note that the real part vanishes above $\eta \approx 1.88\rho_{\text{FEL}}$. This means that the exponential growth stops if the electron energy W exceeds the resonant energy W_r by more than $\Delta W = 1.88\rho_{\text{FEL}} W_r$. The real part of α_2 (*dashed blue curve*) is always negative, hence the eigenfunction $V_2(z)$ drops exponentially. Finally, $\Re(\alpha_3) \equiv 0$, so $V_3(z)$ oscillates along the undulator axis.

$$\begin{aligned} a_1 &= \frac{1}{6}\left(u - \frac{4b^2}{u} - i4b\right), \qquad (10.93)\\ a_2 &= \frac{1}{6}\left(-\frac{1-i\sqrt{3}}{2}u + \frac{2}{1-i\sqrt{3}}\frac{4b^2}{u} - i4b\right),\\ a_3 &= \frac{1}{6}\left(-\frac{1+i\sqrt{3}}{2}u + \frac{2}{1+i\sqrt{3}}\frac{4b^2}{u} - i4b\right), \end{aligned}$$

where the complex function $u(b)$ is defined by

$$u = u(b) = \sqrt[3]{108i - 8ib^3 + 12\sqrt{12b^3 - 81}}. \qquad (10.94)$$

The eigenvalues depend on the relative energy offset $\eta = (W - W_r)/W_r$ through the function $u(b) = u(\eta/\rho_{\text{FEL}})$.

As said above, the maximum value of $2\Re\{\alpha_1(\eta)\}$ is equal to the inverse power gain length. The real parts of the eigenvalues α_1 and α_2, multiplied with $2L_{g0}$, are plotted in Fig. 10.6 as a function of η/ρ_{FEL}. The third eigenvalue α_3 remains purely imaginary even for $\eta \neq 0$. The physical relevance of this figure is discussed in Chap. 6. The maximum value of the growth rate function $f_{\text{gr}}(\eta) = 2\Re(\alpha_1(\eta))L_{g0}$ is unity which means that in this special case the power gain length L_g is indeed identical with the 1D power gain length L_{g0} of a mono-energetic beam, as expected.

Now we make an approximation which is important for computing the FEL bandwidth curve, compare Sect. 5.22. In the vicinity of the maximum at $\eta = 0$ the growth rate function can be approximated by a parabola (second-order Taylor expansion)

$$f_{\rm gr}(\eta) = 2\Re(\alpha_1(\eta))L_{\rm g0} \approx \left(1 - \frac{\eta^2}{9\rho_{\rm FEL}^2}\right). \tag{10.95}$$

The imaginary part of α_1 is needed when ones wants to compute the FEL group velocity, which is done in Sect. 6.5. For this purpose the derivative of $\Im(\alpha_1)$ with respect to the light frequency ω has to be calculated. For this purpose, we express the detuning parameter η in terms of the relative frequency deviation, see Eq. (5.8):

$$\eta = \eta(\omega) = -\frac{\omega - \omega_r}{2\omega_r}, \quad b = b(\omega) = -\frac{\omega - \omega_r}{2\omega_r \rho_{\rm FEL}} .$$

In the vicinity of $\omega = \omega_r$ the derivative of $\Im(\alpha_1)$ is approximately given by

$$\frac{d\Im(\alpha_1)}{d\omega} \approx \left[\frac{d\Im(\alpha_1)}{db}\frac{db}{d\omega}\right]_{\omega=\omega_r} = \frac{\Gamma}{3\omega_r \rho_{\rm FEL}} = \frac{2k_u}{3\omega_r}. \tag{10.96}$$

In order to prove this relation one takes the derivative of Eq. (10.92) with respect to $b = \eta/\rho_{\rm FEL}$, yielding

$$\left(\frac{da}{db}\right)_{b=0} = -\frac{2i}{3},$$

and makes use of $\alpha_1 = a_1\Gamma$ and $\eta = -(\omega - \omega_r)/(2\omega_r)$.

10.4.3 Cylindrical Beam with Constant Charge Density

For simplicity we restrict ourselves to the case of a cylindrical beam with homogeneous charge density. More general beam cross sections are considered in Ref. [10]. Cylindrical coordinates (r, ϕ, z) with $r = \sqrt{x^2 + y^2}$ are used and azimuthal symmetry is assumed. Then $A = A(r)$ and

$$G(r) = 1/A_b \quad \text{for } r \le r_b, \quad G = 0 \text{ for } r > r_b,$$

where r_b is the beam radius. Equation (10.88) reads now

$$A'' + \frac{A'}{r} + i2k_\ell \alpha A = \frac{-2k_\ell \Gamma^3 CGA_b}{1 + k_p^2 CGA_b} A. \tag{10.97}$$

Defining coefficients $u = u(\alpha)$ and $v = v(\alpha)$ by

$$u^2 = \frac{2k_\ell \Gamma^3 C(\alpha)}{1 + k_p^2 C(\alpha)} + i2k_\ell \alpha, \quad v^2 = -i2k_\ell \alpha,$$

the differential equation for $A(r)$ can be written as

$$\begin{aligned} A'' + \frac{A'}{r} + u^2 A = 0 \quad \text{for } r \leq r_b, \\ A'' + \frac{A'}{r} - v^2 A = 0 \quad \text{for } r > r_b. \end{aligned} \tag{10.98}$$

The solutions which remain finite at $r = 0$ and tend to zero for $r \to \infty$ are

$$\begin{aligned} A(r) = c_1 J_0(u\, r) \quad \text{for } r \leq r_b, \\ A(r) = c_2 K_0(v\, r) \quad \text{for } r > r_b. \end{aligned} \tag{10.99}$$

Here J_0 and K_0 are Bessel functions. The continuity of A and A' at $r = r_b$ leads to the system of equations

$$\begin{aligned} c_1 \cdot J_0(u r_b) = c_2 \cdot K_0(v r_b), \\ c_1 \cdot u\, J_0'(u r_b) = c_2 \cdot v\, K_0'(v r_b). \end{aligned}$$

In order to have a nontrivial solution for the coefficients c_1 and c_2 the determinant must vanish

$$det = \begin{vmatrix} J_0(u r_b) & K_0(v r_b) \\ u J_0'(u r_b) & v K_0'(v r_b) \end{vmatrix} = 0. \tag{10.100}$$

The roots $u = u(\alpha)$, $v = v(\alpha)$ of the equation $det = 0$ can be found numerically. Note that a whole sequence of roots (u_n, v_n) of the equation $det = 0$ exists owing to the oscillatory nature of the Bessel function $J_0(x)$. The first root (u_1, v_1) yields an eigenvalue α_1 with a larger positive real part than obtained for the higher roots. This eigenvalue leads to the fastest growth rate and is therefore the most relevant one for the FEL.

It is worthwhile mentioning that a sequence of eigenvalues with positive real part is also obtained for realistic electron beam profiles. The FEL beam can then be characterized in good approximation by the Gaussian modes described in Sect. 10.3. The fundamental TEM_{00} mode has the highest growth rate and dominates in an FEL equipped with a long undulator while higher modes such as TEM_{10} or TEM_{20} have a lower growth rate.

10.4.4 When is the 1D Theory Applicable?

In order to investigate the range of validity of the 1D theory the simplest model of a cylindrical electron beam with constant charge density is too unrealistic. Instead we consider the most important practical case that the beam has a Gaussian transverse charge distribution. So we assume an amplitude depending on $r = \sqrt{x^2 + y^2}$ in the form

$$A(r) = A_0 \exp\left(-\frac{r^2}{2\sigma_r^2}\right)$$

which has cylindrical symmetry, i.e. $\sigma_x = \sigma_y = \sigma_r$. The left-hand side of Eq. (10.88) becomes

$$\left[\nabla_\perp^2 + 2ik_\ell\alpha\right]A(\boldsymbol{r}_\perp). \tag{10.101}$$

The magnitude of the derivative is bounded

$$\left|\nabla_\perp^2 A(r)\right| = \left|A'' + \frac{A'}{r}\right| \leq \frac{2}{\sigma_r^2}A_0.$$

To find the magnitude of the second term in (10.101) we insert for α the eigenvalue α_1 of the exponentially growing solution which is roughly given by Eq. (4.49):

$$\alpha_1 \approx (i + \sqrt{3})\Gamma/2, \quad |\alpha_1| \approx \Gamma = (\sqrt{3}L_{g0})^{-1}.$$

The derivative term can be disregarded if

$$\frac{2}{\sigma_r^2} \ll 2k_\ell\Gamma.$$

It follows that the rms width of the electron beam has to obey the inequality

$$\sigma_r \gg 0.52\sqrt{L_{g0}\lambda_\ell}.$$

As a rule of thumb one can say that the 1D FEL theory is adequate if the rms radial width of the electron beam is much larger than the geometric mean of FEL wavelength and power gain length

$$\sigma_r \gg \sqrt{L_{g0}\lambda_\ell}. \tag{10.102}$$

With reasonable parameters this criterion cannot be fulfilled at ultraviolet and X-ray free-electron lasers. The obvious conclusion is that a realistic description of these high-gain FELs must be based on a 3D theory.

10.5 Statistical Methods and Tools

10.5.1 Current Modulation Resulting from Shot Noise

The quantization of charge is the deeper reason why *shot noise* (in German *Schrotrauschen*) is observed in electronic circuits. This type of noise occurs also in the electron beam of an FEL. Our discussion of shot noise follows the lucid presentation in Ref. [8].

10.5.1.1 Power in Time and Frequency Domain

Consider a real function $A(t)$, for example the current through an ohmic resistor of $R = 1\ \Omega$, and its Fourier transform $a(\omega)$. They are related by

$$A(t) = \frac{1}{2\pi}\int_{-\infty}^{\infty} a(\omega)\exp(-i\,\omega t)d\omega, \tag{10.103}$$

$$a(\omega) = \int_{-\infty}^{\infty} A(t)\exp(i\omega t)dt. \tag{10.104}$$

Since $A(t)$ is real it follows

$$a(\omega) = a^*(-\omega). \tag{10.105}$$

In practice the measurement time is restricted to a finite duration T. If $A(t)$ is put to zero outside the range $-T/2 \le t \le T/2$ we can define

$$a_T(\omega) = \int_{-T/2}^{T/2} A(t)\exp(i\,\omega t)dt \tag{10.106}$$

which will approach $a(\omega)$ for a sufficiently long time T. The average power is

$$\begin{aligned} P &= R\frac{1}{T}\int_{-T/2}^{T/2} |A(t)|^2\,dt \\ &= \frac{R}{4\pi^2 T}\int_{-T/2}^{T/2}\left[\int_{-\infty}^{\infty} a_T(\omega)\exp(-i\omega t)d\omega \int_{-\infty}^{\infty} a_T^*(\omega')\exp(i\,\omega' t)d\omega'\right]dt. \end{aligned}$$

The factor $R = 1\ \Omega$ will be dropped in the following. Interchanging the order of integration and using the approximation

$$\int_{-T/2}^{T/2} \exp(i\,[\omega' - \omega]t)dt \approx \int_{-\infty}^{\infty} \exp(i[\omega' - \omega]t)dt = 2\pi\delta(\omega' - \omega),$$

which is valid for large T, we obtain

$$P = \frac{1}{2\pi T} \int_{-\infty}^{\infty} |a_T(\omega)|^2\, d\omega.$$

This can be written as an integral over positive frequencies

$$P = \frac{1}{\pi T} \int_{0}^{\infty} |a_T(\omega)|^2\, d\omega. \tag{10.107}$$

If we define the spectral density function by

$$S(\omega) = \lim_{T\to\infty} \frac{1}{\pi T}\ |a_T(\omega)|^2 \tag{10.108}$$

then $S(\omega)d\omega$ is the average power within the frequency range $[\omega, \omega + d\omega]$, and

$$P = \int_{0}^{\infty} S(\omega)d\omega \tag{10.109}$$

is the total power.

10.5.1.2 Shot Noise

We consider relativistic electrons ($v \to c$) that are randomly distributed along the bunch and call N the number of electrons in the time interval T. Then $\dot{N} = N/T$ is the average number of electrons per unit time. The absolute magnitude of the dc beam current is

$$I_0 = e\dot{N}.$$

The current seen by a stationary observer has a time dependence

$$I(t) = e \sum_{j=1}^{N} \delta(t - t_j) \quad \text{with } t, t_j \in [-T/2, T/2], \tag{10.110}$$

where the delta functions account for the point-like nature of the electrons in the beam. The average of $I(t)$ over a large sample of identical systems is equal to I_0:

$$\langle I(t) \rangle = e\,N/T \equiv I_0. \tag{10.111}$$

The Fourier transform of $I(t)$ is

$$i_T(\omega) = \int_{-T/2}^{T/2} I(t) \exp(i\omega t) dt = e \sum_{j=1}^{N} \exp(i\omega t_j). \tag{10.112}$$

Using Eq. (10.108) the spectral density function is computed by forming the expression

$$S(\omega) = \lim_{T\to\infty} \frac{1}{\pi T} |i_T(\omega)|^2 = \lim_{T\to\infty} \frac{e^2}{\pi T} \left[N + \sum_j \sum_{k\neq j} \exp(i\omega[t_j - t_k]) \right]. \tag{10.113}$$

When we take the average over a large sample of identical systems, the double sum in this expression vanishes because the times t_j and t_k are random and independent. Moreover, $N = \dot{N}T$, hence we obtain for the ensemble-averaged spectral density function

$$S(\omega) = \frac{eI_0}{\pi}. \tag{10.114}$$

Owing to the delta-function like shape of the current pulses of single electrons this spectral density is independent of frequency. This is a characteristic feature of white noise: the spectral density does not depend on frequency, and for different samples the phase of the current $i_T(\omega)$ in Eq. (10.112) is uniformly distributed between 0 and 2π.

The rms current is defined by $P = RI_{\rm rms}^2$. For white noise we have to limit the bandwidth either by a band-pass filter or—in our case—by the FEL bandwidth $\Delta\omega$. Therefore

$$I_{\rm rms}^2 = S(\omega)\Delta\omega = \frac{eI_0}{\pi}\Delta\omega.$$

The spectral current density is defined by the relation

$$|A_b J(\omega)|^2 = S(\omega) \tag{10.115}$$

where A_b is the cross sectional area of the electron beam. The rms modulation current density $\tilde{j}_1$ needed in Sect. 7.1 is given by

$$\tilde{j}_1 = \frac{\sqrt{I_{rms}^2}}{A_b} = \sqrt{\frac{eI_0\Delta\omega}{\pi}}\frac{1}{A_b}. \tag{10.116}$$

10.5.2 The Gamma Distribution

In this section we study the statistical properties of SASE radiation and demonstrate that the FEL pulse energy fluctuates according to the gamma distribution. The one-dimensional approximation is applied, neglecting the dependencies on the transverse coordinates and treating the electromagnetic waves as truncated plane waves. The number of undulator periods N_u is considered as large so that the frequency ω_ℓ of undulator radiation is restricted to a narrow range. We consider two limiting cases:

(1) The electron bunch is short compared to the length $l_{\text{coh}} = c\tau_{\text{coh}}$ of the optical wavetrains that are produced when the electrons emit undulator radiation. Then the amplitudes of the wavetrains from different electrons must be added and we obtain a single wave packet which is often called a longitudinal mode. In this case the statistical properties of the FEL radiation energy are described by the negative exponential distribution.
(2) The bunch length is large compared to the length l_{coh} of the optical wavetrains. The wavetrains emitted by electrons of sufficient spatial separation do not overlap. In that case several independent wave packets will be excited, and the total FEL pulse energy obeys the gamma distribution.

10.5.2.1 Single Wave Packet

We consider first an electron bunch which is shorter than the optical wave trains. The electrons are randomly distributed along the bunch axis. Particle j emits a radiation field which we write approximately as a horizontally polarized plane wave

$$E_j(t) = E_0 \exp(-i\omega_\ell t)\exp(i\phi_j).$$

The total electric field generated by all electrons in the bunch is

$$E(t) = E_0 \exp(-i\omega_\ell t)\sum_j \exp(i\phi_j). \tag{10.117}$$

The time-averaged field energy is

$$U \propto E_0^2 \left|\sum_j \exp(i\phi_j)\right|^2. \tag{10.118}$$

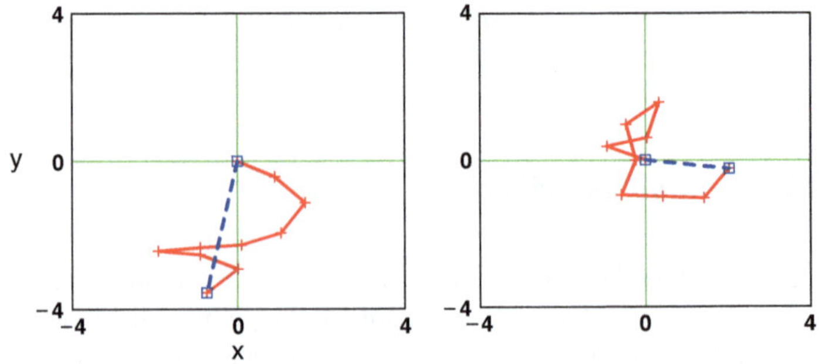

Fig. 10.7 Two examples of random walks in a plane. The step size is 1, the number of steps is $N = 10$. The start and end points of the walks are indicated by *blue boxes*.

We are interested in the statistical fluctuations of the energy. To this end, it is useful to realize that the sum of the phase factors $\sum_j \exp(i\phi_j)$ with random phases can be interpreted as a random walk in the complex $(x + iy)$ plane with unit step size, writing the phase factors in the form $\exp(i\phi_j) = x_j + iy_j$ with $x_j^2 + y_j^2 = 1$. Figure 10.7 illustrates such random walks for two different sets of random phases.

After N steps, the end point of a random walk starting from the origin will have a certain probability to be found in the intervals $[x, x+dx]$ and $[y, y+dy]$. According to the Central Limit Theorem, this probability is given by the two-dimensional Gaussian (see e.g. [11, 12])

$$p(x, y)dxdy = \frac{1}{2\pi\sigma^2} \exp\left(-\frac{x^2 + y^2}{2\sigma^2}\right) dxdy \quad \text{with } 2\sigma^2 = N. \tag{10.119}$$

The distance r from the origin is $r = \sqrt{x^2 + y^2}$. Using $dxdy = 2\pi r dr$ the probability of finding the end point within a distance interval $[r, r + dr]$ is

$$p(r)dr = \frac{r}{\sigma^2} \exp\left(-\frac{r^2}{2\sigma^2}\right) dr. \tag{10.120}$$

The mean square radius is

$$\left\langle r^2 \right\rangle = \int_0^\infty r^2 p(r)dr = 2\sigma^2 = N. \tag{10.121}$$

The dimensionless variable

$$\xi = \frac{r^2}{\left\langle r^2 \right\rangle}$$

obeys the simple exponential probability distribution

$$p(\xi)d\xi = \exp(-\xi)\, d\xi, \qquad (10.122)$$

where we have used $d\xi = (2r/\langle r^2 \rangle)dr = (r/\sigma^2)dr$. Since the radiated field energy U is proportional to the square of the electric field we can conclude from this equation that the probability distribution of the radiation field energy is given by

$$p_1(u)du = \exp(-u)du \quad \text{with } u = \frac{U_1}{\langle U_1 \rangle}. \qquad (10.123)$$

Here the index 1 indicates the presence of one wave packet. This is the negative exponential distribution. The most probable value of the FEL pulse energy is zero.

10.5.2.2 Many Wave Packets

When the bunch is far longer than the optical wave trains there will be more than one, say M, wave packets. Each of them will obey the exponential distribution (10.123). For sufficient spatial separation these M "modes" are uncorrelated and the respective probability distributions are statistically independent. We want to prove that the probability distribution for the total energy of M independent wave packets is given by the expression

$$p_M(u)du = \frac{u^{M-1}}{\Gamma(M)} \exp(-u)du \qquad (10.124)$$

with the gamma function $\Gamma(M)$. The proof is made by complete induction. The statement is true for $M = 1$, then Eq. (10.124) is identical with (10.123). Now assume that formula (10.124) has been proved up to some value of M. We make the step from M to $M + 1$:

$$p_{M+1}(u) = \int_0^u p_M(v) \cdot p_1(u - v)dv$$

where we make explicit use of the statistical independence of the wave packets by taking the product of the probability distributions p_M and p_1. Inserting formula (10.124) we get

$$p_{M+1}(u) = \int_0^u \frac{v^{M-1}}{\Gamma(M)} \exp(-v) \cdot \exp(-(u - v))dv = \frac{u^M}{\Gamma(M + 1)} \exp(-u).$$

This completes the proof.

Formula (10.124) is the gamma distribution as defined in the mathematical literature. It is not directly applicable at the SASE FEL but must be modified slightly. From the above derivation it is obvious that the variable u is the ratio of the total FEL

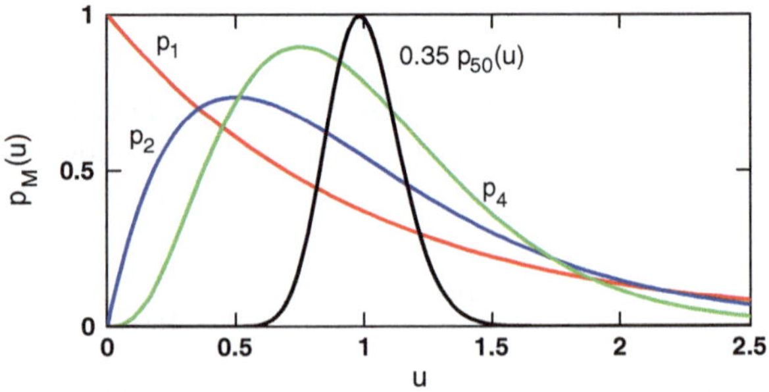

Fig. 10.8 The gamma distribution for $M = 1$ (negative exponential distribution) and $M = 2, 4$. For the large value $M = 50$ the distribution is scaled down with the factor 0.35. The area under each of the curves is equal to 1, see Eq. (10.126).

pulse energy U_M (summed over all M wave packets), divided by the average energy of a single wave packet. However, the energy contained in a single wave packet is difficult to measure, while its average value is just the measurable total average energy, divided by the number of wave packets:

$$\langle U_1 \rangle = \langle U_M \rangle / M.$$

Hence it is convenient to introduce a modified dimensionless variable $\tilde{u}$

$$\tilde{u} = \frac{U_M}{\langle U_M \rangle} = \frac{u}{M}.$$

Then the gamma distribution can be written in terms of measurable quantities:

$$p_M(\tilde{u})d\tilde{u} = \frac{M^M \tilde{u}^{M-1}}{\Gamma(M)} \exp(-M\tilde{u})d\tilde{u} \quad \text{with } \tilde{u} = \frac{U_{\rm rad}}{\langle U_{\rm rad} \rangle}.$$

Here $U_{\rm rad} \equiv U_M$ is the energy of the entire radiation pulse.

To simplify notation we drop the "tilde" in the following and write the gamma distribution in the form

$$\boxed{p_M(u)du = \frac{M^M u^{M-1}}{\Gamma(M)} \exp(-Mu)du \quad \text{with } u = \frac{U_{\rm rad}}{\langle U_{\rm rad} \rangle}.} \tag{10.125}$$

The gamma distribution for $M = 1, 2, 4, 50$ is plotted in Fig. 10.8. For very large M it approaches a Gaussian centered at $u = 1$. The distribution is normalized to unity

$$\int_0^\infty p_M(u)du = 1. \tag{10.126}$$

The mean value and the variance of the normalized FEL pulse energy are

$$\langle u\rangle = \int_0^\infty u p_M(u) du = 1, \quad \sigma_u^2 = \left\langle (u - \langle u\rangle)^2 \right\rangle = \frac{1}{M}. \tag{10.127}$$

It must be pointed out that only two limiting cases have been considered here: (1) fully overlapping wave trains, yielding a single wave packet (or optical mode), and (2) several perfectly separated modes. In reality partially overlapping wave packets will exist too. One can approximate their probability distribution by a gamma distribution with a non-integer index M. This is done in Chap. 7.

10.6 Conventions and Frequently Used Symbols

In this book we use the international system of units (SI). Specifically

$c = 299792458\,\text{m/s}$: speed of light in vacuum,
$e = 1.60218 \cdot 10^{-19}\,\text{A s}$: elementary charge,
$\varepsilon_0 = 8.854 \cdot 10^{-12}\,\text{A s/(V m)}$: permittivity of free space,
$\mu_0 = 4\pi \cdot 10^{-7}\,\text{V s/(A m)}$: permeability of free space,
$m_e = 9.109 \cdot 10^{-31}\,\text{kg}$: rest mass of electron,
$\hbar = h/(2\pi) = 1.05 \cdot 10^{-34}\,\text{J s}$: Planck's constant.

The charge of the electron is written as $q_e = -e$. The electron energy in the accelerator is quoted in MeV or GeV = 1000 MeV as usual (1 MeV = $1.602 \cdot 10^{-13}$ J). A right-handed Cartesian coordinate system (x, y, z) is used. The electron beam moves along the z direction, x is the horizontal displacement of an electron from the nominal orbit and y is the vertical displacement. Following the usage in many modern physics textbooks we call the field $\boldsymbol{B}$ the magnetic field because it appears in the Lorentz force and is thus directly responsible for the deflection and focusing of the electron beam ($\boldsymbol{H}$ is called the "magnetizing field"). We will often call $\omega = 2\pi f$ the frequency although it is an angular frequency.

In the following Table 10.1 we summarize the designation of frequently used quantities, their dimension in SI units, their meaning, and the equation or the chapter where the quantity is introduced.

Table 10.1 List of frequently used symbols

Symbol	SI Units	Meaning	Eq./Chap.
$\boldsymbol{A}$	V s/m	Magnetic vector potential	Sect. 10.1
A_b	m^2	Electron beam cross section	Chap. 5
$\alpha_1, \alpha_2, \alpha_3$	m^{-1}	Eigenvalues of third-order equation	Eq. (4.49)
$\boldsymbol{B}$	T	Magnetic field vector	–
B_0	T	Peak magnetic field of undulator	Chap. 2
$\beta = v/c$	–	Normalized velocity of electron	–
$\bar{\beta}$	–	Averaged normalized velocity	Eq. (2.11)
$\beta_x(z), \beta_y(z)$	m	Horizontal/vertical beta function	Sect. 6.3.1
β_{av}	m	Average beta function	Eq. (6.12)
$\varepsilon_x, \varepsilon_y, \varepsilon$	m	Emittance of electron beam	Eq. (6.11)
ε_n	m	Normalized emittance	Eq. (6.13)
$\tilde{E}_x$	V/m	Field of light wave	Eqs. (4.1), (4.10)
$\tilde{E}_z$	V/m	Space charge field	Eq. (4.18)
η	–	Relative energy deviation	Eq. (3.14)
$F_0(\eta), \tilde{f}_1(\eta, z)$	–	Particle distribution function	Eq. (10.40)
φ_E	–	Phase of field $\tilde{E}_x$	Eq. (5.21)
φ_{j1}	–	Phase of current $\tilde{j}_1$	Eq. (5.22)

(continued)

Table 10.1 (continued)

Symbol	SI Units	Meaning	Eq./Chap.
G	–	Gain function of FEL	Eqs. (3.26), (5.3)
γ	–	Lorentz factor of electron	Eq. (1.2)
γ_r	–	Resonant value of Lorentz factor	Eq. (3.13)
Γ	m^{-1}	Gain parameter	Eq. (4.44)
I_0	A	DC electron beam current	–
I_{peak}	A	Peak electron beam current	Chap. 5
j_0	$\mathrm{A/m^2}$	DC electron beam current density	Eq. (4.4)
$\tilde{j}_1$	$\mathrm{A/m^2}$	Modulated current density	Eq. (4.4)
K	–	Undulator parameter	Eq. (2.8)
$\widehat{K}$	–	Modified undulator parameter	Eq. (3.30)
k_ℓ	m^{-1}	Wave number of FEL radiation	Eq. (3.1)
k_p	m^{-1}	Space charge parameter	Eq. (4.44)
k_u	m^{-1}	Undulator wave number	Chap. 2
λ_ℓ	m	Wavelength of undulator/FEL radiation	Eqs. (2.19), (3.9)
λ_u	m	Undulator period	Chap. 2
L_{g0}	m	1D power gain length	Eq. (4.51)
L_g	m	3D power gain length	Chap. 6
L_{sat}	m	Saturation length	Eq. (7.5)
$p_M(u)$	–	Gamma distribution	Sect. 10.6
P_{beam}	W	Electron beam power	Eq. (5.16)
P_{rad}	W	Undulator radiation power	Eq. (2.21)
P_{sat}	W	FEL saturation power	Eq. (6.27)
ψ	–	Ponderomotive phase	Eq. (3.7)
ψ_b	–	Bucket center phase	Eq. (5.23)
ψ_m	–	Microbunch phase	Eq. (5.24)
r_b	m	Radius of electron bunch	Chap. 6
ρ	$\mathrm{C/m^3}$	Electron beam charge density	Eq. (4.3)
$\tilde{\rho}_1$	$\mathrm{C/m^3}$	Modulated charge density	Eq. (4.3)
ρ_{FEL}	–	FEL parameter	Eq. (4.47)
σ_η	–	Normalized RMS energy spread	Chap. 6
σ_r	m	RMS radius of electron beam	Chap. 6
$\bar{v}_z$	m/s	Longitudinal speed	Eq. (2.11)
W	MeV	Relativistic electron energy	Chap. 2
W_r	MeV	Resonant value of electron energy	Chap. 3
$w(z)$	m	Radial width of photon beam	Eq. (6.19), Sect. 10.3
X_d	–	Diffraction parameter	Eq. (6.21)
X_ε	–	Angular spread parameter	Eq. (6.21)
X_γ	–	Energy spread parameter	Eq. (6.21)
ζ	m	Internal bunch coordinate	Eq. (3.12)
z_R	m	Rayleigh length	Eq. (6.18), Sect. 10.3

References

1. L.D. Landau, E.M. Lifshitz, *Course of Theoretical Physics, Volume 1: Mechanics* (Butterworth-Heinemann, Boston, 2003)
2. H. Goldstein, *Classical Mechanics* (Addison-Wesley, Reading, 1959)
3. J.B. Murphy, C. Pellegrini, *Introduction to the Physics of the Free Electron Laser, Laser Handbook*, vol. 6 (North Holland, Amsterdam, 1990), p. 11
4. J.D. Jackson, *Classical Electrodynamics*, 3rd edn. (Wiley, New York, 1999)
5. Z. Huang, K.-J. Kim, Review of X-ray free-electron laser theory. Phys. Rev. ST Accel. Beams **10**, 034801 (2007)
6. D.A. Edwards, M.J. Syphers, *An Introduction to the Physics of High Energy Accelerators* (Wiley, New York, 1993)
7. A.M. Kondratenko, E.L. Saldin, Generation of coherent radiation by a relativistic electron beam in an ondulator. Part. Accel. **10**, 207 (1980)
8. A. Yariv, *Optical Electronics in Modern Communications* (Oxford University Press, New York, 1997)
9. D. Meschede, *Optik, Licht und Laser* (Teubner, Stuttgart, 1999)
10. E.L. Saldin, E.A. Schneidmiller, M.V. Yurkov, *The Physics of Free Electron Lasers* (Springer, Berlin, 2000)
11. F. Reif, *Fundamentals of Statistical and Thermal Physics* (McGraw Hill, New York, 1965)
12. E. Lohrmann, V. Blobel, *Statistische und numerische Methoden der Datenanalyse* (Teubner Studienbücher, 1998)

Index

A
Angular spread
    parameter, 95

B
Bandwidth of FEL, 68
Beta function, 90
Betatron oscillation, 92
Bose-Einstein statistics, 5
Brightness, 165
Bucket
    FEL bucket, 34, 77, 81
    RF bucket, 34
Bunch compression, 142

C
Canonical momentum, 186
Coherence
    longitudinal, 117
    transverse, 115
Coherence time, 120
Coherent transition radiation, 152
Collective effects, 145
Correlation function, 119
Coupled first-order equations
    non-periodic model, 202
    periodic model, 51

D
Detuning, 66, 86
    energy, 63
    parameter, 66
Diffraction, 94
    parameter, 95
Diffraction limit, 166
Distribution function, 193
Doppler effect, 3, 16

E
Echo-enabled harmonic generation, 128
Eigenfunctions, 57
Eigenvalues, 57, 212, 214
Electro-optic effect, 150
Emittance, 90, 207
    criterion, 93
    measurement, 158
    normalized, 91
    slice emittance, 158
Energy spread, 87, 92, 179
    parameter, 95
Exponential gain regime, 72
Exponential growth regime, 1

F
FAST, 85
FEL
    amplifier, 25
    oscillator, 25
    seeding, 25
FEL parameter, 56, 65
FEL radiation
    wavelength, 28
FEL-seeding, 123
FERMI at Elettra, 126
Fixpoint, 33
FLASH, 133
Focusing

P. Schmüser et al., *Free-Electron Lasers in the Ultraviolet and X-Ray Regime*,
Springer Tracts in Modern Physics 258, DOI: 10.1007/978-3-319-04081-3,

natural, 90
strong, 90
weak, 90

G
Gain function
high-gain FEL, 64
low-gain FEL, 35
Gain guiding, 94
Gain length
1D case, 57
3D case, 95
Gain parameter, 56
Gamma distribution, 120, 221
Gaussian beam optics, 94, 204
GENESIS, 84, 85
GINGER, 85
Group velocity, 100
Growth rate function, 86

H
Hamiltonian
low-gain FEL, 32
non-relativistic, 183
pendulum, 184
relativistic, 185
Helical undulator, 60, 188
Helicity, 60, 191
High-gain harmonic generation, 125
High-gain regime, 65
Higher harmonics, 20, 36, 114

I
Integro-differential equation, 196
Internal bunch coordinate, 30

L
Larmor formula, 15
Laser
free-electron laser, 7
quantum laser, 5
Laser amplifier, 30
LCLS, 168
Lethargy regime, 59, 64
Linear regime, 1, 73
Lineshape, 19
Liouville theorem, 195
Lorentz factor, 3
Lorentz transformation, 15, 16, 137
Low-gain limit, 64, 197

M
Madey theorem, 35, 65, 66, 199
Microbunch, 40, 75, 77, 113
Mode
Gaussian, 204
longitudinal, 120
transverse, 115
Mode competition, 115

N
Nonlinear regime, 75

O
Optical eigenmode, 5
Optical resonator, 5
Optical transition radiation, 113

P
Pendulum equations, 31
Phase space
high-gain FEL, 76
longitudinal, 193
low-gain FEL, 33
pendulum, 184
Phase velocity, 99
Photocathode, 135
Photon quantum state, 5
Pierce parameter, 56
Plasma frequency, 56
Polarization
circular, 61, 191
linear, 17, 190
Ponderomotive phase, 28
Power
FEL saturation, 73, 97
SASE FEL, 110
synchrotron radiation, 3
undulator radiation, 17

Q
Quality factor
cavity, 141
Quantum diffusion, 179
Quasi-monochromatic radiation, 117

R
Rayleigh length, 94, 205
Relative energy deviation, 31
Resonance energy, 30

S
SACLA, 168
SASE
  fluctuations, 117, 120
SASE FEL, 67, 107
  power, 109
Saturation, 71
  FEL power, 73
  length, 109
Seeding, 66
  by higher-order harmonics , 124
  echo-enabled harmonic generation, 128
  high-gain harmonic generation, 125
Self-seeding, 173
Separatrix, 33, 76, 185
Shot noise, 218
Slippage effects, 101
Space charge, 87, 136, 145
  parameter, 56
Space charge field, 47
Spontaneous emission, 6, 7
Stimulated emission, 5
Superconducting cavity, 139
Superposition principle, 73
Superradiance, 105
Surface resistance
  superconductor, 141
SVA approximation, 45
Synchrotron radiation, 3
  coherent, 145
  power, 3, 18

T
Third-order equation, 55, 197
Transmission function, 110
Transversely deflecting structure, 149

U
Undulator
  magnet, 3, 12, 134
  period, 3, 12
  tapering, 176
Undulator parameter, 4, 13
  modified, 38, 48
Undulator radiation, 3
  angular width, 20
  higher harmonics, 20
  lineshape, 19
  power, 18
  spectral energy, 21
  wavelength, 17, 21

V
Vector potential, 186
Vlasov equation, 195

W
Wake fields, 147
Water window, 114

X
X-ray beam lines, 179

MIX
Papier aus verantwortungsvollen Quellen
Paper from responsible sources
FSC® C105338

If you have any concerns about our products,
you can contact us on
ProductSafety@springernature.com

In case Publisher is established outside the EU,
the EU authorized representative is:
Springer Nature Customer Service Center GmbH
Europaplatz 3, 69115 Heidelberg, Germany

Printed by Libri Plureos GmbH
in Hamburg, Germany